Commonsense Engineering: Simple, Effective Practices for Engineering and Quality Control

Gary L. Casey PE

Acknowledgements

I would like to especially thank two people, Mark Miller and Marcos Nassar for their expert constructive criticism and most of all their enthusiastic encouragement. Without that I would probably not have even started on this adventure. My late father, Wayne Casey encouraged my curiosity into all things and my mother, Mildred Casey still provides the stable view of the world and keeps me in my place. Everyone remembers a teacher in their formative years - let's not forget my seventh-grade teacher, Ms. French, for sparking my search for knowledge. And finally, my wife Sarah had the patience to tolerate all the time the preparation of this book required.

Table of Contents

Introduction

Anyone who has worked in a position involving some kind of repetitive task has certainly thought about how to improve the performance of that task. The task could be developing and manufacturing a product, serving a customer, transporting goods, or whatever. Virtually any process can respond favorably to improved control and tracking, and many products purport to help you do just that. How well they do it, and how to see through some of their weaknesses, is the subject of this book.

My intent in writing this book is to describe logical and common-sense approaches to solving typical problems in product launch, manufacturing system design, and quality control. While it may sound like this is only about the manufacturing of goods, the principles apply to a wide range of activities. The idea is to reduce an activity to its simplest and most effective elements and therefore make it practical for even the smallest companies to apply the principles and thrive in today's competitive markets. Therefore, it is a "practical guide." As such, however, it is not an all-inclusive manual to replace all others. In cases where I feel there isn't anything to add (or anything to complain about!), I leave the practitioner to use another tool. Or they can use their own common sense—often the best method of all.

This book discusses the following topics:

Statistical properties: strengths and weaknesses. Practically all statistical methods are based on the assumption that the population can be described by the normal distribution curve, which is sometimes misleading.

C_{pk}: When it works and when it doesn't. This widely used term is often underutilized, misused and over-emphasized. I try to show how it can be effectively used, when to be suspicious, and how to manipulate it.

Capability: what it is and how to use it. "Capability" is an overall measure of the ability of a product or process to accomplish what is intended. It is not a binary evaluation; something cannot be considered to be either capable or incapable, but rather *how* capable. The process (or product) capability index, or C_{pk}, is perhaps the most common measure of capability, but it has weaknesses and by itself is not all-inclusive.

Control: In or Out. What does it mean when someone claims their process is "in control?" What is the appropriate response when a process goes "out of control?" I discuss the appropriate use of this tool.

Significant characteristics and why they are insignificant. Almost all organizations seem to cling to the concept of "significant" or "critical" characteristics as if their life depended on it. They need to get over it, and this topic explains how.

Sample-inspection procedures. Organizations often decide to inspect a small sample as a way to confirm quality, but seldom is that implemented rationally. A typical sample-inspection procedure is often a "poster child" for incorrect methods. Seeing the weakness of the conventional approach of sample inspection should open the reader's eyes to other misguided quality procedures.

Quality systems as a business and what to watch out for. Any number of presumably well-meaning people have proposed quality systems. How do you pick the best one—or better yet, implement the best parts of all without any of the distractions? A difficult but possible job.

A guide to product development procedures. I propose a method of implementing a product or project development procedure that doesn't get bogged down in details, unnecessary reviews, and red tape.

Design reviews and how to avoid their pitfalls, which are often due to the differences in perspective the reviewer and reviewee. I propose a rational and commonsense approach.

Specifications: what do they specify? Specifications are at best a summary of the almost infinite number of characteristics of the actual product. Thus, the objective has to be simply to minimize the short-comings and recognize that it's not possible to eliminate them entirely.

Design and process failure mode and effects analysis (FMEA). The phrase brings fear and dread to the typical engineer. But it doesn't necessarily have to—let's bring a little reason (and, dare I say, common sense) to the subject.

A configuration management system is something that is sooner or later required in every company. Similar to the constitution and laws of a government, the configuration management system can provide the framework and guidance, as well as the record-keeping for the company. Some systems work well, some don't, and why some systems are preferred, at least by me. Let's make a system that is powerful, simple, and easy to use.

Single-piece flow as a manufacturing concept. This is a tidbit that mostly applies to the repetitive serial processing of a product. It is included to give an example of how a creative approach to a process can result in big improvements.

Efficiency versus effectiveness: what's the difference? People often get the two confused, and as a result their efforts are often less effective—and less efficient—than what is possible.

Importance versus urgency and how to tell the difference. These two terms are also used frequently and often confused with each other when trying to prioritize activities. It is critical to be able to tell the difference between them and act accordingly. When managing multiple concurrent activities gets tough, not knowing the difference can be truly destructive.

The natural order of things, or how to recognize that there might be a logical order of doing things, hopefully intuitive, but not always. Let's sharpen our senses and look for the natural order.

This book is not all-inclusive. Nor is it thorough in all things it does cover—to accomplish that would require volumes. It does describe an overall approach that I believe captures the essence of commonsense practices without burdening the user with unnecessary or marginal activities. It essentially follows the 80–20 rule: the intent is to describe 20 percent of the possible efforts that will get you 80 percent of the way to the final objective. The objective? To create a company that accomplishes its goals in a rational, effective, and efficient manner. In other words, to use ordinary common sense!

Statistical Properties: Strengths and Weaknesses

It is probably best to always keep in mind the principal uses of statistical properties and statistical analysis: they are simply ways to summarize the properties of a population (more than one item, property, person, etc.) using numbers.

Statistical analysis can't accurately predict a characteristic of a single item, but it can predict the likelihood of the property being within a given range. Say the average age of a group of two people is forty. We may have no clue as to the age of each person, but we can say it is more likely that the age of each person is closer to forty than to some other age. If someone asked me to guess the age of one of them, would I guess ninety? No, the place to start is forty, even though that is probably not the exact age of either one.

To come to some common ground in describing a population, we need some definitions. Certainly *average* is one that everyone understands—simply add up the characteristic of each one in the population and divide by the number in the population. For example, the average, or mean, income in the United States might be $100,000. Wait, you say, not very many people make that much. Correct, but there are a few who make a lot more. That's why we often use the term *median*—the number at which an equal number of the population is higher and lower than the value. For example, the US median income might be $40,000, a number that might be more useful because it discounts the outliers—in this case, the very rich.

The next most common tool for describing a population is probably the *normal distribution* function, a concept dreamed up by Frederick Gauss. He concluded that the likelihood of any individual characteristic being close to the mean is greater than that of being farther away, and he created an equation that describes this effect. He further created the term *standard deviation (σ)*, which is the distance from the mean that includes a large percentage of the population—68.2 percent, to be exact. Note that in the figure below, from *Wikipedia*, the portion of the population outside a given deviation from the mean is shown—outside of about 3σ, one can expect to find only about 0.2 percent of the population (0.1 percent on the high side and 0.1 percent on the low).

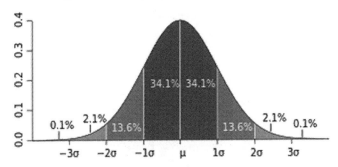

Using the data from a population, the mean and the standard deviation can be calculated—most easily by pushing a button (the "STDEV" function) on a spreadsheet. If you want to get into the details of the calculation, you may do so on your own; however, most people prefer to just push the button.

After the two fundamental numbers—the average and standard deviation—are determined, a wide range of characteristics of the population can be predicted. One is the number of the population that is at a certain distance from the mean. We usually use the number of standard deviations (sigma, as in "6-sigma out") as the measure. This distribution curve uses what I call in this book "linear math" in that it extends equally and infinitely on both sides of the mean. For example, the average age of a population might be forty with a standard deviation of ten. One can now predict the number of people who are at least the age of sixty, or whatever.

But wait—there are other factors at work. The normal distribution prediction would also predict that there is a finite number of people age two hundred, and some are age minus ten. So obviously, other factors distort, or in this case truncate, the normal distribution curve. And therein lies the difficulty.

Almost everyone new to the subject will now say, "All this fuss about normal distribution is OK for most things, but *my* distribution is different. This normal distribution stuff won't work for me." The practical fact of the matter is that most populations conform to the normal distribution well enough to allow its use as the primary tool. Tools have been developed to account for a distribution that is, say, fatter, or even nonsymmetrical. However, 90 percent of the time, a simple normal distribution will work for most purposes as long as you don't stray too far from the mean. Note all the caveats: most of the time, most purposes, not too far from the mean. That means you will normally want to ignore data from outliers—data unusually far from the mean.

Finally, recognize that, as stated above, any statistic is a number that is used to *describe* a population—it isn't the population itself. Quite often we don't have the data for the entire population; if we did, we would have less use for the statistic. Say we have a population of a thousand but we don't want to spend the money to measure all one thousand, so we pick a random sample of one hundred to measure. How well does that describe the population? Pretty well, you say?

But how well? Let's say the standard deviation of the sample is ten, so it is equally probable that the sigma of the entire population is higher or lower than that. But to make it more useful, we can use a principle called the "central limit theorem" and then we can make a statement such as, "I have a ninety percent confidence that the population will have a sigma of less than twelve."

Following is the equation for what's called the "lower confidence limit" (LCL), using a small sample of a population. C_{pk} is the calculated *process capability index* of the sample, n is the sample size, and Z_a is a constant, which happens to be 1.28 for a 90 percent confidence level. (My apologies for momentarily stepping ahead of myself—discussion of C_{pk} is in the next chapter. You can skip ahead and then come back if you like.)

$$C_{pk} - Z_a \sqrt{\frac{1}{9n} + \frac{C_{pk}^2}{2n-2}}$$

The central limit theorem says the more samples you have, the more likely the data is to match the data of the entire population (another way of saying the more you gamble at Vegas, the more likely you are to lose). Some will say there is a sample size (I believe it is thirty-three, but who cares?) that is "statistically significant" and that sampling fewer parts is not valid. Hogwash. Any sample size greater than one is significant, but more samples produce more significance. You can find discussions of the central limit theorem in most statistics books.

As a practical example of the use of statistics, consider shafts that are machined to a specific diameter on a lathe. If normal manufacturing processes are used, most shafts will likely be very close to the mean and the variation can be described quite well using the normal distribution model. However, every once in a while, the tool will break and one shaft will be radically different from the others.

Should we include the diameter of this shaft in the population? This is where common sense comes in; the answer is "probably not." The reason is that the diameter of this shaft is different because of a defined cause—tool breakage—while the rest of the population has diameters that are produced by "normal" variation in the process. Why do we care? Because we will likely want to mount an effort to reduce (diameter) variation, and if the broken-tool shaft is included, the variation we are looking for will be obscured. In actual practice, we might start two projects—one to reduce the "normal" variation and one to reduce tool breakage.

So my recommendation is to keep all data in the population until you have some reason to believe it should be excluded. It probably means you will start with data that looks like a mess until you exclude some of the extraneous data (some will tag it with the label "special cause data"). But don't just forget about it; find the specific cause and work to eliminate it. Often the outlier data will be binary in nature (either the tool breaks or it doesn't) and not the result of variation. In summary, you might as well assume that your population will behave according to the normal distribution rule, but then look for reasons to exclude some of the data—look for defined special causes.

Here's an actual example of this: We were grinding thin diaphragms to final thickness and then measuring the flatness: zero is flat, plus was bowed up (convex), and minus was bowed down. The standard deviation didn't look promising, but on careful examination we found that the actual distribution was bimodal in that it had two peaks, with the peaks virtually equal and equally spaced from zero. This pointed to the defined cause of the diaphragms being bowed: they were being temporarily flattened during the grinding process, only to pop back into shape (well, out of shape) afterward. We were measuring diaphragms randomly, sometimes bowed up and sometimes bowed down, and that obscured the data. So we went to work on both causes: the bow in the diaphragm and reducing the variation in grinding, which was actually quite small. One thing we did not take into account was that we could have found a way to use the diaphragms in a single orientation (bowed up or bowed down). We might have found that a bow in the right direction would improve performance. Thus, we missed a chance to find a new opportunity for improvement, which we would have found if only we had followed the statistics to their logical conclusion.

In summary, most populations seem to "want" to be normal and as a consequence can be analyzed assuming they conform to a normal distribution. However, most any population will exhibit some deviation from normal and often there is "gold" in that deviation. So besides trying to reduce variation, also look at the deviations from a normal distribution. You'll be glad you did!

C_{pk}: When It Works and When It Doesn't

The term "C_{pk}" seems to be so widely used it is perhaps worth the effort to examine where it came from, what it is, where it works, and where it doesn't.

The capability (C) of the process (p) index (k) is a statistical measure of how many standard deviations separate the process or product mean from the nearest specification limit. Since there are typically, but not always, both upper and lower specification limits, typical usage often designates three C_{pk} values: the upper C_{pk} (UC_{pk}), the lower C_{pk} (LC_{pk}), and finally the lesser of the two—simply C_{pk}. The upper C_{pk} is the separation of the product or process from the upper spec limit, and the lower one...well, you get the idea. When just C_{pk} is specified, it is assumed to be the lesser of the UC_{pk} and LC_{pk}. While the person who reports C_{pk} values may ignore which C_{pk} is the limiting one, the user—the one who has to "fix it"—usually cares a great deal.

Below is a typical picture of a product- characteristic distribution compared with the upper and lower spec limits. Crude? Certainly! I hope you get the idea that such things are rarely neat and tidy and almost never look like the examples in textbooks.

Somewhere along the line, someone decided that a simple number of standard deviations was, you might guess, too straightforward

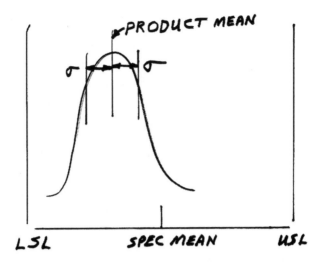

(too much common sense?), so they decided to divide by three. So, if there are four standard deviations between the process mean and the nearest spec limit, the C_{pk} is reported as being 1.33 (much more esoteric, yes?). And to further complicate things, someone decided that there are two more types of C_{pk}: the preliminary process capability and the mature process capability. The preliminary one, which some define as P_{pk}, is a measure of the initial process and is presumably not yet affected by the aging (wear and tear and other as-yet-unknown factors) of the process. So where did this decision to divide by three come from? Certainly the idea of determining the capability in terms of the number of standard deviations from the spec limit is a fundamental concept, but the number three was truly just invented for no real reason. Purists will disagree and come up with all kinds of esoteric arguments to justify their position. The best approach is to just use the process capability index as a tool and not get too wrapped up in the justification.

Always remember that numbers are simply surrogates for reality—they themselves are not reality. So the usefulness of any number—C_{pk} in this case—is only as good as how well the number represents the actual parameter (the reality) we are interested in. We'll discuss this in more detail later.

Any number of organizations have put forth "requirements" for the C$_{pk}$ of processes and products. Anything that has a specification can be described by a C$_{pk.}$ The larger the C$_{pk}$ value is, the better are the parts or process compared with the specification. Often the required C$_{pk}$ is 1.33 (4 sigma). Someone picked what he or she thought was a tolerable number of parts outside the spec limit, and the result was 1.33 (no, not 1.3 or 1.5, but 1.33). Then for a preliminary C$_{pk}$ (P$_{pk}$), 1.67 (5 sigma) was picked, presumably for some rational reason. *The point is that any number picked as a requirement is just that—a number selected from any one of the numbers that could have been used.*

This is all logical if the data for all members of a population are known. But what if only a sample is measured? Then we can use the central limit theorem to predict the lower confidence level, as discussed earlier. We can solve for the required C$_{pk}$ of the sample knowing the LCL of the population. (As many instructors would say, "That is an exercise left for the student." But you can fool the instructor by using the "solver" function in Excel.) The bottom line is that we need to know the required C$_{pk}$ of the sample in order to state, "I am ninety percent confident that the C$_{pk}$ of the population is at least XXX." Following is a chart that shows the approximate minimum C$_{pk}$ as a function of sample size for a population C$_{pk}$ of 1.33:

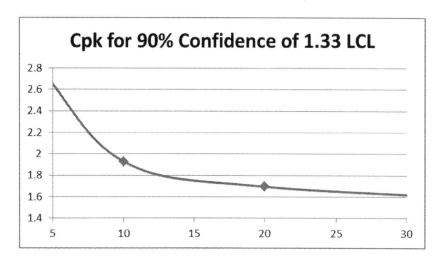

Note that as the sample size is reduced, the required C_{pk} of the sample rises dramatically. For example, if you wanted to use a sample of only 5, the required C_{pk} is 2.46. And as the sample size is increased, the C_{pk} doesn't drop as much as you might like—at 100 the C_{pk} is 1.47. Bottom line: if you want to use a small sample size to "prove" the product meets a 1.33 C_{pk} requirement, the product had better be pretty good indeed! We will talk more about this in the chapter "Sample-Inspection Procedures."

C_{pk} is usually used to measure the capability of a process or product, but sometimes people get confused. Take this true story of a product that was sold: The supplier stated that the C_{pk} of a given characteristic was 1.5. But the last process done was to inspect 100 percent of the product. He found that the C_{pk} was 1.1, and he rejected all parts that were outside the specification. The resulting population then had a calculated C_{pk} of 1.5. He therefore proudly proclaimed to the customer that the product exceeded the required C_{pk} of 1.33 and everything was OK. However, the customer looked at the data and rejected that conclusion because the inspection showed a C_{pk} of 1.1, which was not up to the 1.33 standard. In fact, that was why the 100 percent inspection was implemented and the product as sold *did* meet the 1.33 standard set by the customer (as a result of sorting at inspection)—and therefore why the product was "OK" with a C_{pk} of 1.5. The customer reviewed the data and said, "No, it's not OK, because the real C_{pk} was 1.1."

Who was right? I took the position that the supplier was right, and here's why: The inspection step was a *process* and not a *test*. It changed the population because it threw away bad parts, and therefore it was a process, not just a test. A pure test would merely measure something and report the result, while a process changes something. In this case, the *population* changed characteristics because some parts (the ones out of spec) were removed.

The measure of the *processes* up to that point was 1.1, but the measure of the *product as sold* was 1.5.

Here is a typical—or perhaps not so typical—example to show how a simple statistic like C_{pk} can be used to mislead (or to clarify, depending on whether you are buying or selling). Let's use a characteristic with only a single spec limit to make it simpler. In this example, we'll use a burst-pressure requirement, which would be typical of virtually anything expected to contain a fluid under pressure. Let's pick a spec limit of 400 psi and burst pressures that are close to what might be expected—with a little bit of "artistic license" thrown in to illustrate the point. There are five parts, and the tested burst pressures are 500, 600, 1,000, 399 (just to keep you wondering), and 600 psi. The resulting average burst pressure is 619.8, and the standard deviation is 228.3, giving a C_{pk} of 0.32. Pretty pathetic capability—and one of the five parts doesn't even meet the spec, although by only 1 psi. What would have happened if that part had burst at 400? Then all parts would have met the spec, and in the bad old days (before the dawn of statistical enlightenment), we might have pronounced the test a success and the product ready to sell. Now I think we know better—the C_{pk} would have not been acceptable.

But let's take a closer look at what the C_{pk} value implies. Remember from our previous discussion of statistics that C_{pk} calculations utilize "linear math" and attempt to predict performance "everywhere"—all the way from $-\infty$ to $+\infty$. If we use the tail of the normal distribution curve, we can predict from the above data that a full 1,500 ppm will burst at pressures below zero! Hey, I didn't make it up—I just used standard statistical methods. Of course, it's physically impossible for the burst pressure to be negative, so in this case, using C_{pk} and its assumed standard distribution curve as a surrogate of reality is certainly flawed.

Now let's look at the same data from a different perspective. The specification in this case is a measure of part strength—greater strength leads to a higher value for burst pressure. But what if we used "weakness" as the characteristic? That is certainly a valid concept, is it not? After all, error, which is another word for weakness, is probably the most typical characteristic used in the industry, and the better the performance of the part, the lower the error. So let's invert all the values in the example; thus 1/burst = weakness. Do the same math, and the C_{pk} is now 0.44—still not good, but 38 percent higher than before. That's a big difference. Which is correct: 0.32 or 0.44? Let's see if weakness is a better stand-in for reality. Now we can predict that no parts will fail at pressures below zero, a much better result than before, when we predicted that .15 percent would fail with no pressure at all applied. But wait—now we predict that 500 ppm will fail at pressures above infinity (a weakness below zero). Hmm. You can see that there is no absolute answer. We could use the log of weakness as the criteria...but I digress.

Let's do a "deep dive" (a favorite command from managers who don't have to do the diving) into the reason for the poor performance. We find that the one part that failed below the spec had a process that was simply skipped (I'm making this up—after all, it's my book). Now let's put in process controls that will absolutely prevent (be suspicious when anyone says that) that occurrence. We retest five more randomly selected parts and find that the burst pressures are 600, 700, 1,000, 700, and 610. Doing a quick calculation of C_{pk} using both burst and weakness as the characteristic, we find that the C_{pk}s are 0.66 and 1.33, respectively. Now what are we to think? Our company policy might say that a C_{pk} of 1.33 is acceptable as proof of capability but 0.66 is not. And we could legitimately say that "weakness" is the more logical requirement, but, as they say, "It depends on whether you are buying or selling." I dare you to do the same math on a typical specification using "accuracy" and "error" as the two characteristics.

Here is yet another look at the same data. Most of the parts failed in a fairly narrow range, between 600 and 700 psi. One part can be considered to be an anomaly, with a burst pressure of 1,000. If that part had failed in the normal range, the C$_{pk}$, regardless of method, would be above 1.33 and the part would have likely been deemed acceptable. Should one part that is dramatically *better* than the average be "used against us" as proof that the process is incapable? Again it depends on whether you are buying or selling. If the C$_{pk}$ is being used as evidence that the parts will function acceptably, then yes, the parts are capable. But if you are using the C$_{pk}$ value to evaluate the process (not the part), then you should conclude that the process is *not* capable, as there is too much variation from the mean.

Now let's make up something (again). We do a "failure" analysis on the very strong part. Remember, it should be considered a failure because it was the cause of the process not being capable. We find that the part indeed failed, but in the process of failing, the two components jammed together instead of blowing apart. If I fix the characteristic that allowed that to happen, I now have both a process and a part that is fully capable. Done. Or is it? The part that failed at 1,000 psi could be considered to have failed from a special cause and therefore should be excluded from the population, whether or not that special cause could be prevented. Perhaps that would be more of a commonsense approach. Since the special cause created an outlier on the "good" side, eliminating it from the population would do no harm and would allow us to better evaluate the true variation in burst pressure.

Bottom line: C$_{pk}$ is certainly a powerful indicator of conformance to a requirement, but *only* when used with intelligence, a healthy degree of skepticism, and, most of all, common sense.

Capability: What It Is and How to Use It

Capability is a term that is much talked about but seldom rigorously implemented. It might appropriately be described as the confidence level by which a process or part will meet the requirements. Statistical tools are generally used for determining capability, and the most-used tool (rightfully so, I believe) is the process capability index (C_{pk}) previously described. The time to start using statistical tools to evaluate a design is as soon as the second part is made. But actually, it could be done even earlier—in the design process, since manufacturing variation can be simulated with modern analysis tools quite accurately. But always be aware that the criteria for the evaluation of capability are usually determined by humans. It may be logical, but it could just as well be completely arbitrary. Beware!

For the word *capability* to be used in reference to the ongoing quality of a process or product, multiple criteria should be satisfied. First, the product or process must be unchanged and have historic data. If a product has been deemed "capable" but is then subjected to a change in design or process, capability has to be reestablished. Generally, the history has to be consistent for some period—say three to five production lots, three months of continuous experience, or the like. Second, performance has to fall within the requirements by some accepted margin; typically this will be confirmed by a C_{pk} of 1.33 to 1.67.

As is often the case, you can expect a comment like, "That makes sense for a high-volume product manufacturer, but I build just a few aerospace parts at a time. It doesn't apply to me." A valid retort might be, "How many screws, rivets, or other fasteners do you install?" It would likely be very useful to apply the principle of capability to the fastener installation process. And the commenter shouldn't be surprised if asked, "What is the torque spec that is typically required? Is the torque driver capable of meeting that?" The likely response is, "Of course it does." "Well, prove it." That's what capability is all about.

What is capability not? Mostly, it is not proof positive that a product or process will meet the requirements. It is merely an arbitrary (OK, maybe not completely arbitrary) criterion that can be used as a way to communicate between people. Instead of saying, "All of our processes are robust," one should perhaps say, "All of our processes are capable to our criteria." Minor difference? The point is that capability isn't an absolute, but a relative, term.

So what is the value of this term? How do you use the process capability index to improve your operation? Perhaps the biggest value is that it allows you to summarize the performance of a number of processes or products and to prioritize your improvement efforts. Start by improving the least capable process, and go from there.

And if you ask someone, "Are all these process capable?" and you get the reply, "Of course," be on the lookout—something surely isn't right!

Control: In or Out?

Control as a term related to a process or product is a useful concept and should be considered by anyone. Often I'll propose that an accepted practice is something to be suspicious of, but in other cases, a useful tool is often ignored by many. Control is one such tool.

Control describes the performance of a product or process compared with historical performance, and as such it has nothing to do with specification limits. Most quality manuals provide the formulas for calculation of C_{pk} control limits, so I won't duplicate them here. Suffice it to say that the "rule," as documented in most control manuals, essentially says that if the mean or variation (standard deviation) of the process or product is observed to be outside historically established control limits, something must be done.

For example, suppose a process stays well within the specification limits, but suddenly the mean shifts. The product is still well within the spec limit, so why should anything be done? Because it changed—that's why! Murphy's law states that any change is usually a bad change. Our ultimate goal is to understand the process, and any significant change should be addressed. Perhaps the change resulted from a component characteristic's change and was predicted—in that case, the out-of-control condition could be written off and closed with no corrective action. But maybe the process could be adjusted to bring the mean back to where it was, and that should perhaps be done.

But what happens if the process mean originally deviated from nominal (but was still in spec) and it suddenly became closer to the target? The process got better, not worse, so nothing has to be done—right? Wrong! If the change is studied and understood, perhaps it can be made permanent. Otherwise we run the risk that it will shift again, and the next time it may not be so pleasant.

I recall glancing through production records and finding that the product produced one day was dramatically better than normal—these were great parts! I looked and found that yes, indeed, the process was documented as being "out of control," but nothing was done. I was beside myself when I found that the parts, clearly the most valuable ones we had built, had already been shipped. Here was evidence that the product could be built to a much higher level of quality, but the evidence was gone. When I confronted the manufacturing manager, he was puzzled. He said, "But the product got better, not worse. Why should we have done anything?" Needless to say, an education session ensued.

Here's another point: since the existence of control as a process characteristic doesn't depend on spec limits, it can be effectively employed when there are no spec limits. For example, say you are managing a call center and are tracking the results of your staff. You have no real basis for the establishment of spec limits, so you are operating "in the blind, Right?" Wrong—you can employ *control* as a tool. If the performance of one operator is outside the control limits, something should be done. If performance is below the lower control limit, the operator needs to be retrained or perhaps terminated. If an operator is performing above the upper control limit, that operator needs to be studied to find his or her technique. An operator is within the control limits? Spend your time elsewhere.

In summary, the use of control limits can be very useful and should be considered for any process or product. As you continually address out-of-control conditions, the control limits will gradually tighten up and the outcome will progressively improve. And that, after all, is the real goal.

What's Significant about Significant Characteristics?

Nothing gets a rise out of me faster than a discussion of "significant characteristics." After all, if some characteristics (requirements) are "significant," then by definition the remaining requirements are "insignificant." Or at least "less significant." No customer wants to hear a trusted supplier say, "No, we don't meet that requirement, but after all, it's insignificant." There is no such thing as a real significant characteristic—after all, all of them are significant, or the customer wouldn't have put them in the specification to begin with.

The following discussion will attempt to point out the fallacy of using significant characteristics as a control method. As such, it flies in the face of procedures that have been developed by a number of respected quality organizations. Furthermore, if you take the following position, you will certainly anger your customer and at the least spark a long, colorful debate.

The creation of the term *significant characteristic* (SC) and the related terms *critical characteristic* (CC) and *safety critical characteristic* (SCC) probably grew out of the limitations of paper and two-dimensional thinking to adequately describe a product. For example, a mechanical product will likely be described by a three-view drawing with dimensions for each feature. Each dimension may have a specific tolerance or may be defined by "block" tolerances, which are defaults for dimensions with a specific number of figures after the decimal point. Does a simple dimension and tolerance adequately describe the feature? Perhaps not, as discussed in the "Specifications: What Do They Specify?" chapter. Can other descriptors be added? Perhaps the designer is unsure and, as a result, is worried about the performance of the product. In response, it is tempting to add an SC tag to the feature, which in essence says, "I'm not sure what to do about this, but I'm worried that the product will fail if this feature isn't controlled better than the others."

So an SC, a CC, or an SCC designator is an admission of a shortfall of the designer—not necessarily of the design. Remember, if one feature is "significant," then the others must be "insignificant" or at least "less significant." Really? Doesn't the designer want all the design requirements to be met? Since, in the context of significant characteristics, it is not OK to allow the SC requirement to fail or not be met, it must logically be OK to allow the others *to* fail. Proponents will say, "But if *this* feature is out of specification, then the product will fail, and I can't allow that." That implies (no, it actually explicitly states) that other features *can* be out of specification without product failure. Logic, or common sense, indicates that the very fact that some features are designated significant and others aren't implies that the design imposes unnecessarily tight requirements on some features and not on others. *Why not selectively adjust the requirements on all features until they are equally "significant?"*

To know what to do about SC, CC, or SCC designations, one has to delve into the company's quality manual or, if not there, publications of an appropriate quality-system purveyor. Often, a vague requirement that the feature "must be under statistical process control" (SPC) will be found. What does that mean? Most people seem to accept the statistical tracking of the characteristic as proof of SPC. This usually consists of a chart of C_{pk} or some other statistical descriptor. But what happened to the "C" in SPC? Without implementing the control part, it isn't really SPC, is it? There are many examples of C_{pk} charts delivered with products as "proof" of SPC, where the C_{pk} has not been compromised for years. Why bother? Is that a good use of resources? In many cases, however, the statistical characteristic will be tracked and used to control the process, just as it should be. But what about the other product characteristics? The usual answer is, "They do not require SPC because they are not SCs." By now, this should be starting to sound like nonsense.

So how did the SC, CC, or SCC designator get applied to some features and not others? It generally happens during the design phase and is done by the designer or the internal or external customer. And it is likely done because someone's judgment says the particular characteristic is particularly important to the function of the product. Or someone thinks some failure modes are more important than others, so the associated characteristics must be carefully monitored. Often company policy (the quality manual, typically) will state that customers' designations must be carried through to the component level, and then, because the designation is a contracted requirement, it isn't negotiable.

Is it possible to predict in advance which characteristics are more likely to cause problems or fail to meet requirements? What if the manufacturing process changes—hopefully increasing but maybe decreasing the robustness of the characteristic? What happens if the customer changes the usage of the product? In that case the specification should be revised. And what if 100 percent inspection is incorporated, making a failure of that characteristic almost impossible? What about those other "insignificant" characteristics— is nobody watching them? All these questions make the incorporation of the SC designation anything but common sense.

You can see that the whole SC concept is based not on reality but on assumptions and judgments. Our objective should be to make sense of it all and accomplish the design task without designating some features as more significant than others. It all starts with a design that is as robust as possible. For mechanical-part design, this will likely result in individual tolerances being applied to most, if not all, characteristics. Is it not logical (common sense) that each requirement, which exists for a specific and individual purpose, will require an individual tolerance?

For example, a typical electrical connector may have a sealing surface on the inside and an outside that interfaces with no other component. The sealing surface likely has a number of precise requirements to ensure it performs its function. The outside, however, is only there to "contain" the inside. Perhaps only a minimum value for the wall thickness is required. The manufacturer could then make it any size larger, with no limit. Most designers will balk at this and say, "I don't want them to make it really big," or "It's a molded part—they can meet 0.xxx tolerance." But why make the tolerance based only on the perceived ability of the assumed manufacturing process? The manufacturer won't make it too large, as it will cost the company money; or too much plastic will cause shrinkage and make the internal surface difficult to control. The designer might then (correctly, but needlessly) say, "I need to control the thickness to minimize shrinkage and control the internal shape." But that is *already* controlled by appropriate tolerances. Using the design as a tool to anticipate processing problems is usually futile, since the designer cannot possibly be aware of all potential processes that might be used.

So the answer about how to control the product in response to requirements hasn't so far been answered. The real answer is, *state the actual requirement for each feature, and require all features to be capable.*

This sounds like all features will now be considered "significant." Exactly—what a concept! But what to do about it should be covered in the quality manual and will *not* be to track SPC data on all features. It *will* require all features to be proven capable, but that is not as big a task as it sounds; refer to the chapter "Capability: What It Is and How to Use It." Since all characteristics will now be significant, the SC, CC, and SCC callouts can all be eliminated.

As an example, let's return to the connector discussed above. In the past, the sealing surface would likely have been designated "significant," so nothing has changed there. The outer surface, previously designated "insignificant," is now important. However, the outer diameter requirement is now gone—the requirement is *not there*. The manufacturer then can easily locate the outside surface such that the internal feature can be capable. For example, a manufacturer could add a dimple in the outside surface to eliminate shrinkage at some spot on the internal surface—it's all up to that person. Now the supplier actually has an *easier* time of it, even though all features are now considered to be significant. Thus, eliminating the significant characteristic protocol can lead to a lower-cost, higher-performance product. Certainly, more burden is placed on the designer, who must now determine the true requirement for each characteristic of the product, and the producer of the product must now control each and every characteristic equally to the level previously required only of "significant characteristics." But once these requirements are met, the resulting product can be expected to perform to a consistently high standard. So how does the designer draw the outer surface? I suggest just drawing the surface as envisioned, but without any dimension attached. It is up to the supplier to decide on the exact shape. Or the supplier could base the design of the die on the solid model sent to him or her, truly using the capability of modern design tools. I've seen plastic parts designed this way with fewer than half the dimensions otherwise required.

I hope you can now appreciate how illogical the concept of significant characteristics really is. They are at best a crutch that is used when the true requirements are not known. It would make more common sense to leave significant characteristics where they belong – anywhere but in the specification for a product.

Sample Inspection Procedures

A number of procedures have been proposed with the intent of qualifying a large group of parts by inspecting only a small fraction. The first question to answer when contemplating this process is, why do sample inspection at all? What is the intent, and what is expected to be accomplished? Quite often the justification is that the practitioner is simply uncomfortable with the assumption that nothing has changed in the processes and thinks that testing a few will soothe those fears. In other words, no rigorous or rational justification is used.

This is, however, not to say that without rational justification, sample inspection should never be used—best available (common-sense) judgment is often irrational but still justified. Perhaps one doesn't quite trust a supplier even though that supplier has formally demonstrated compliance with the requirements. This might be the most often-used criteria, and in such a case sample inspection could provide an early (or, more accurately, "barely in time") warning that something has changed.

The most common practice is to pull a number of parts from a lot (population) and inspect them against some criteria. This is usually not intended to be an inspection of all requirements, but instead a small number of selected requirements—often mistakenly labeled significant characteristics (SCs). You can already see that normal statistical justification criteria are being compromised—a small number of parts and a small number of requirements for each part are inspected. Then what are the criteria by which the lot is accepted or rejected? Quite often, if all inspected characteristics meet requirements (are in spec), the lot is passed. Is this justified? And what has become of the lot? In many cases the lot is quarantined during the inspection process. But occasionally the lot will already have gone to the next step; in that case what, if anything, does the inspection accomplish?

Let's assume that reasonable statistical monitoring of all processes is already in place. How do you then implement a robust sample-inspection process? Refer to the earlier discussion of the typical statistical criterion—the C_{pk}. Then apply the central limit theorem, which is based on the idea that any sampling of a large population can only predict the characteristics of the total population with a certain confidence level—never 100 percent. So, what confidence level is required of the lot? One can pick any number, but a logical value might be 90 percent. And what specifically does one want to be confident of? Again, a logical answer might be that the lot must have a C_{pk} of a certain value, typically stated, "I require a 90 percent confidence that the lot has a C_{pk} of 1.33 or greater."

The rest is easy: Pick any sample size—say five—and inspect the parts. Use the central limit theorem to determine the C_{pk} requirement of the sample population. Using approximate numbers, for a sample size of five and a population C_{pk} of 1.33, the required C_{pk} is 2.6. If the sample population exhibits a C_{pk} of at least 2.6, the lot can be predicted to have a C_{pk} of at least 1.33 with a confidence level of 90 percent and therefore would be considered to "pass," based on the central limit theorem.

But what happens if the sample inspection fails to exhibit the required C_{pk}? Simple—and there are really only two choices. One is to reject the lot, but that might be expensive, both because that lot is now lost to production (it is scrap) and because the supplier of (or the customer for) that lot now must absorb the cost of the entire lot. This first choice is usually unacceptable, or at least undesirable.

The second choice is to inspect more parts (randomly selected, of course). The data from the original parts inspected is *not* to be discarded but will be included in the new sample population. It is not acceptable to simply pick another five and keep trying until the lot passes. Let's say five more parts are inspected, bringing the required C_{pk} down to 2.1, due to the larger sample size of ten. If the new, larger population meets that requirement, the lot can be considered to pass. For the new parts added to the inspection population, only those characteristics that didn't originally meet the requirements need be inspected, keeping the cost of the extra inspection to a minimum. All the other characteristics have already passed, and once a "pass" is recorded, no further testing of those characteristics is required.

So what happens if adding more parts doesn't produce positive results? The number of parts to be inspected can be increased yet again—even to the point that 100 percent of the population is inspected—which is one of the acceptable outcomes if the product doesn't meet the stated C_{pk} requirements. In the case of 100 percent inspection, it is usually acceptable for the lot to be considered a "pass" if all parts are in spec, although this criteria should be stated in the company's quality policy manual, not decided on an individual basis—or, for that matter, decided by an individual.

How does one determine which of the possibly many characteristics of a part are subject to sample inspection? Often one might conclude that just "significant characteristics" need to be subject to sample inspection. Wrong! See the chapter "What's Significant about Significant Characteristics?" The characteristics to be included should be determined in the control plan and will likely be better described as "bellwether" characteristics. Such characteristics are determined by analysis of the manufacturing process, not by which ones are thought to be more important than others. For example, if a plastic part has fifty dimensions that are all formed by a similar die or molding process, one of those could be considered a bellwether characteristic. The one with the most stringent tolerance might be selected simply because if that one meets requirements, then the others likely will also. Again, refer to the chapter on significant characteristics.

In summary, sample inspection can never guarantee the acceptability of every part in the population. But carefully done, it can be used to generate reasonable confidence in the performance of the population as a whole. And that's a good thing.

Quality Systems as a Business

In recent and not-so-recent history, any number of products have been presented with the intent to increase the overall quality of products and processes. In earlier times, these products were in the form of books and manuals; but recently they seem to have taken on a different form. The original concept of introducing a "system" that can improve a company is still there, but it has been augmented by a supporting infrastructure. For instance, a system might be created and marketed as the next "good way to do business." Then a family of seminars is created, using instructors who are connected—either working for or certified by the creators—with the intent of spreading the word. Adding yet more might be a system of certifying company employees as being trained in the system. Hey, I'm not knocking it—you may read this book and want to hire me to give a seminar!

What do all these systems have in common? Sure, they all purport to improve the user companies, but lots of products, including a simple book or manual, have that intent. The common thread is that *they all make someone money.* They are products no different from yours, created and marketed in such a way as to make a profit for the creator. All of them, including the international quality certification organizations, are intended to be profitable for the originator. Is that a good reason to avoid them? No, but it is a good reason to be cautious.

At one time GE was promoting a thing called "Six Sigma" and presumably the original thought was that the standard deviation of all processes and products should be less than one-sixth of the specification limits. But one could claim to be a Six Sigma company without necessarily complying—it all depended on the fine print. Several years ago I looked at one of the Big Three automotive companies' Six Sigma display wall. The charts and graphs all touted their successes, but I could not find the symbol "σ" or even the word "sigma" anywhere. They did show how many processes were improved, but the measurement was usually in yield improvement, not variation reduction, although that was implied by the reduction in numbers of failures. The point is that in many instances, the core goal of the system gets lost in the details of the implementation.

A number of purveyed quality systems propose methods of dealing with specific portions of the overall picture, such as the decision-making process. Certainly most of the commercially available quality systems have some advantages. And since nearly all quality systems compete for the same business, it was inevitable that one, such as Shainen, would come along that would claim not to compete but to embody in theirs the essence of all the other quality systems.

Following are some examples of quality systems in action:

The "lean" initiative makes a lot of sense—eliminate waste in the manufacturing system and therefore make it lean. You certainly can't argue with that. But hasn't the intent of manufacturing engineering been to do exactly that for the last hundred years or so? Of course. So what is different? Perhaps it is the rigor by which the process is analyzed. Unfortunately, sometimes the rigor introduces shortsightedness, even though the objective is noble. Often the focus is on the total number of feet (or miles) traveled by a component during assembly. This embodies a certain amount of logic, but is it really *the* important criteria? An example might be a factory with a process common to all products that requires the use of a large, expensive machine. In one company, early studies apparently showed that using one big machine to process all parts was more cost-effective than buying smaller machines, each dedicated to a single product. All parts were then brought to the "Big Machine" for that process, usually staged in batches. The "Lean Team" came into being, added up the miles traveled by the products and brought back the concept of providing a separate Big Machine for each line. Yes, that reduced the miles traveled, but did it reduce the manufacturing cost? Or even reduce the production time? Proponents will argue that the total manufacturing time was indeed reduced, making the company more responsive to the customer. Also true, but did it improve profitability? Could they have more easily and cheaply concentrated on increasing the flexibility of the Big Machine? The answer is very likely a resounding "yes."

Other initiatives are fixated on reducing inventory—certainly a worthy goal, at least at first glance. But what is the true cost of inventory? One could argue that it is only the time value of the money invested. But there is also the risk of that inventory becoming obsolete. I once observed a company that built ahead a large number of finished goods because someone was "sure" the customer was going to launch the program and the company would have a head start supplying the product. But the customer canceled at the eleventh hour, making all the parts in inventory surplus. Much gnashing of teeth and blame throwing ensued, but the parts were allowed to stay in inventory with the thinking that some could be sold for different purposes—to me a logical decision. But a year later, in a lean frenzy, the parts were thrown out, with the perpetrator bragging that a million dollars' worth of inventory had been eliminated. OK, but the risk of the parts becoming obsolete was zero—they already were. There was no interest to be paid on the money it took to build them, as that was already a sunk cost. So the only cost of retaining the parts was the cost of the space they took up in the warehouse—a very small number. And the potential income from their sale was lost forever. A smart move? I think not.

But how do the quality systems relate to increased effectiveness as a company? Sometimes the effectiveness increases, but sometimes not. The company with the Big Machine might have been providing an improvement in response to customer orders, in which case instant response would be very valuable. Or it may have been building a product in response to a very stable demand, in which case responsiveness has a low value. Or maybe both. If the latter were the case, then perhaps the centrally located Big Machine could have been simply better utilized, processing urgent products quickly and processing other products as time allowed. Don't you suppose that would have been a more effective approach than to duplicate the Big Machine several times over?

Are all parameters equally valued? Of course not. The distance traveled by a small electronic component from initial fabrication to final sale is probably measured in thousands of miles. Should that be valued the same as the distance traveled by a large locomotive casting? Not likely. Should all inventory be valued solely by the book cost? Probably not. Inventory in the middle of the process in a fast-response company may be very expensive—but maybe not. Example: A company supplies product on order, and most products require a common subassembly or a number of them. Doesn't it make sense to keep a supply of these subassemblies in stock so the final product can be built quickly when ordered? On the other hand, a product could have been overbuilt, resulting in a large final-assembly inventory. Not good; but after the fact, what is one to do about it? Storing the final product might cost next to nothing, but the finance types could urge that the parts be thrown away to "get them off the books." Why?

All the quality initiatives that have been proposed and have drifted in and out of vogue must have had some benefit—otherwise it's not likely they would have been successful. Let's take a look at some of them:

One of the first I was exposed to was the "Kepner-Tregoe Decision Analysis" methodology. Its basic idea was to rationalize the decision-making process, a result you certainly can't argue with. Various methods were proposed. Some were a little "hokey," but many were quite useful. One was a priority-ranking system that, among other things, could be used to convert a customer's stated desire to a product-feature priority ranking. Usually one item in the list of desires is always "low cost." Wrong, said Kepner-Tregoe. Low cost isn't one thing—it is everything. In other words, it is the common denominator used to evaluate all desired features. Examples: How much are customers willing to pay for a reduction in failure rate from .01 to .001? How much are they willing to pay for a one-pound weight reduction? How much up-front money are they willing to pay for a one-dollar production cost reduction? This analysis continues on with all the other features. By using cost as the basis for comparison, a rational ranking can be generated. The core concept is, "Money isn't one thing—it's everything."

Another Kepner-Tregoe concept is that of pushing customer-desire ranking back through the process to the source. Let's say the dominant customer desire is reliability. Which feature of the product most contributes to the lack of reliability? (In other words, which has the most potential to *improve* reliability?) Which characteristic of that feature has the most potential? Which process step in building that feature has the most potential? Which feature of that process step...? Push the analysis back as far as possible. It might be that changing the shape of the button on the machine that makes the widget that goes into the component that turns the lever that...

OK, you get the idea. The end result, or at least the hope, is that large, complex problems can be progressively prioritized and broken down to the source—a very useful and powerful concept.

Another early quality initiative is Total Quality Management (TQM). The idea here is similar to Kepner-Tregoe in that to improve the quality of a part or process, you continuously work on the largest contributor to the problem. One case involved improving the highest-volume product in our company. A number of teams were created, and my team was tasked with improving the yield of the calibration process—the process that resulted in the highest dollar amount of scrap since the part was almost fully assembled by the time it was calibrated. The yield was running 92 percent, and the boss told me I really blew it when I picked that process, that the calibration yield couldn't be improved because of the "nature of the beast." The plan the team developed was to address not necessarily the largest problem first, but problems that could be improved at the least cost and in the least time. A priority system was developed. Sound a little like a Kepner-Tregoe analysis? One of the top items was something called "failure to trim." The visual detection system in the laser trimmer looked for a characteristic on the circuit board so it could establish a position. If it was unable to do that, it rejected the part as "failure to trim." As it turned out, the reason for this was that the focus of the pattern-recognition system could drift. Yes, the team could have installed better optics, but instead we established a daily refocus process and instantly improved the yield by 1 percent. See where this is going? Another interesting change was triggered by looking at the pattern-recognition methodology: it used an L-shaped black resistor pattern, and the system moved until it found one leg of the L and then moved on the other axis to find the other leg. Guess what—the original programmer set it to look for the short leg of the L first, and if it missed, the part was rejected. We reversed the process to look for the long leg first and virtually eliminated the "failure to trim" losses. After a few months, the yield was consistently running over 99 percent, when we were initially told that 92 percent was the "nature of the beast."

The Six Sigma system also had the potential to improve profitability, mostly of manufacturing systems. The idea was to reduce variation of all systems from any source. That's OK, but not all variation contributes to poor product quality, so some intelligent prioritization has to take place. To do that, the specification limits of all product and process characteristics have to be rational—certainly a tall order. One thing a Six Sigma program can highlight is that not only product but also process characteristics need specifications. Quite often in the "bad old days," the process yield was used as the measure of the process. That violates a cardinal rule (well, my cardinal rule anyway): "Don't ever use the *product* to measure the *process*." If the process has its own specifications, it can be evaluated against those without waiting for product failures.

The point is that all these systems have useful features, and the challenge is to pick the gold from each, not necessarily embrace all their tenets. The rigorous use of common sense is especially productive here.

Product Development Procedures

The first question to ask about product development procedures might be whether standardized product development and launch procedures are a good idea. There are numerous examples of successful product launches that didn't use any formal procedures. The product development process is all about creativity, so won't a standardized process stifle creativity? But a standardized process can eliminate oversights and prevent wasted effort. It also can allow those with less capability or experience to perform the task with fewer errors. As with most things, there are two sides to the argument, and the correct answer is probably a compromise that encourages the best of each point of view.

Simplicity is the overriding virtue of any standardized approach. One automotive OEM has a standardized development procedure for transmissions described in written manuals of several thousand pages. Surely even a large crew of trained monkeys could implement the procedure. But is it cost-effective? Are the practitioners spending more time and effort concentrating on meeting the letter of the complex "law" than effectively implementing the intent of the project? My experience indicates that is likely to be the case.

Even though a rigid procedure may sound desirable, the problem is that the focus often shifts from the real objective toward meeting the procedural requirements. For example, an elaborate multistep launch procedure was once developed at a company that will remain anonymous. At the end of each step, a review by peer groups was required and the results scored. The scoring was part of the project manager's performance review, and the project couldn't continue to the next phase unless the score was "acceptable." You can see that getting a good score got the immediate attention of any "victim" of this procedure. While each of the line items by itself was logical and useful, the implementation of the procedure was focused on the score obtained by the engineer rather than the success of the product. For instance, one required item read, "Was there a risk assessment presented?" Regardless of whether the risk assessment showed the project was viable or whether the assessment itself was even accurate, the score was based on whether it was merely presented—easy to get a good score on that one.

Another procedural weakness becomes evident when the review is done by a multifunctional group and the requirement for continuation is consensus. Any member is then able to "veto" the positive vote of all the others. But some groups are inherently configured to be the "watchdogs" of the organization—specifically, those typically called the Quality Department. The group may have a perfectly valid position that their area of responsibility is being compromised by the project and therefore will give it a "no" vote. What to do? In my experience the dissenting group (in this case, Quality) is usually pressured into changing its position based on other considerations—business, marketing, or whatever. That is the *wrong* approach. Upper management's inherent responsibility is to resolve these issues and override subordinate decisions when appropriate—not to force them to change their position. The Quality representative might say, "The capability of this product doesn't meet our objectives as stated in the Quality Manual. Therefore I cannot approve it." Management might then (perhaps correctly) state, "The marketing opportunity justifies taking this risk.

Thanks for pointing it out; and maintain your position, but we will go ahead anyway." In that way the failure remains in the documentation for future "lessons learned." The integrity of the dissenting group is maintained, and everyone knows the status of the project and, most importantly, understands the reason for continuing.

Many (probably most) members of upper management, in my experience, are reluctant to exercise their responsibilities and find it easier to command subordinates to change their position—exactly the wrong thing to do. But what if the subordinate consistently refuses to approve projects and forces management to become involved? It could be just to avoid any implied or potential responsibility in case of project failure. This is a matter to be dealt with between the manager and subordinate in a performance review— possibly the subordinate's last!

Any procedure should be based on real requirements, not desires. A procedure, almost by definition, must then state the *minimum* requirement. If it always states the author's idea of utopia, it may never be met. And how can the practitioners be expected to meet a series of these utopian and often mutually exclusive targets? Should the product be of the lowest possible cost? Of course. And exhibit the highest possible performance? Of course. But to write both into the procedure makes no sense since neither can actually be met. So what should the line items require? That the project meets the initial expectations. Exceeding them is obviously desirable, but remember, the procedure should be a list of *requirements*, not desires. This cannot be emphasized enough.

What, then, is the logical structure of a project? It should probably be in reference to that universal measure—money. The project should have an approved budget to reach a given position by a given time. At the end of each phase, the progress should be evaluated against the initial objectives in view of the time and money spent. Continuation to the next phase will then be approved—or not. This requires that the reviewing management level empowered to spend the money is actually involved in the review process. This in turn requires that the process be simplified as much as possible, or the person at that level of management will lose interest (they have short attention spans, you know). Project-review meetings have been known to last for days, often with no empowered management in sight. Of what use is that? The product of the review should be the approval (or not) to proceed with the next phase—and to spend the required money.

Most projects can be structured using the following four project phases:

1. Concept Development. The engineering department evaluates the product requirements and the various approaches to satisfying same. Individual feature or component performance requirements are determined. If the customer (or management) has not created a complete product-performance document, now is the time to do so. A specific concept is then proposed that is projected to meet all requirements. The review at the end of this phase will result in approval to proceed to the hardware stage.

2. Engineering Development. A design is created by combining appropriate features found viable in the previous phase. Validation testing is directed at the individual features, but not necessarily at the complete product. A complete drawing (or definition) package is created, which usually means a complete, released drawing package will exist at the end of this

phase. Approval of this phase means the design is authorized for continuation; for this to be meaningful, a documented design (with fully released drawings) must exist. Note that this is in *direct contradiction* to conventional thinking, which assumes that by the end of this stage, only experimental (unreleased) drawings will exist. But how can you expect a crowd of people to "approve" a design when it can later be changed with only superficial approval? That makes no sense—but that's the usual procedure.

3. Design Validation. The prototype product will be fabricated in accordance with the approved design. This will be validated in reference to the original product requirements document. Any changes in design that arise from test failures must go through a repeated design-validation testing procedure. At the end of this phase, the performance of the parts built to the previously approved design will be reviewed. The approval might be as-reviewed, or it may be conditional, with certain changes approved. The authorization will presumably be to go ahead with production tooling and capitalization.

4. Process Validation (often called Product Validation). The complete production process will be implemented, including all capital equipment. All production tooling will be procured, and parts will be built using the equipment and tested according to the original requirements. The end result of this phase is the part in production, since final production approval was essentially given at the end of the previous phase.

Each phase has a substantial capital requirement, generally increasing with each one. The phases are divided according to defined spending blocks so that management can evaluate progress against cost and approve the next phase's expenses.

Note that phase 4 includes production approval (given at the end of phase 3) with the condition that final validation testing is successful. Presumably, no review needs to take place after the final process-validation testing is complete. After all, why would the expenses of production tooling and process validation be approved if production weren't tacitly approved?

That said, one more approval might need to be made. Logically, the signature level required should be for the total monetary "exposure" of an item. Let's say that phase 3 costs were estimated to be $100,000. Someone with the authority to spend $100,000 should approve the initiation of phase 3, even though the expenditures will be made in $10,000 increments. After the overall phase is approved, approval of the smaller steps should be automatic—no need to go "back to the well" each time a purchase is initiated.

This is an important concept. Too often a program is "approved" without specific authorization for the cost involved. The unfortunate soul assigned the project will now find him or herself trying to justify each expenditure, usually to a lower-level manager who may or may not be "on board" the program. Think how easy it would be if the engineer knew that each incremental expense would be "automatically" approved—and, of course, charged against the initial authorization.

So who, then, is the "approver" for each line item? Let's say the phase is approved for $100,000 in expenses. The manager of the program is then authorized to spend how he or she sees fit without having to go back for approval to spend $10,000, even if he or she theoretically isn't approved for that level of expense. Note that the program manager could spend the $100,000 for a Tahiti vacation if he or she likes, and it would work—until his or her next and probably last performance review. My point is that a manager should be given the responsibility for a program and—here's a novel and commonsense idea—given the tools to execute the program.

A short comment on expenditure authorization limits. Often over the course of history, authorization limits are adjusted during times of duress (budget tightening) such that the approval limit at every management level is reduced. The result is that a given expenditure has to be approved at a higher and higher level, usually rising to a management level that essentially "doesn't care" about that amount of money—an application of the Peter Principle to purchase authorization. What happens is that all the lower management levels simply sign the request, since they know they can't approve the expense anyway. The request finally works its way up to a management level that likely doesn't even know what it's all about and certainly doesn't care. Then the approval (disapproval) is likely to be totally arbitrary. Instead of the monetary control being tightened, it is actually loosened. Anyone who works in a large company can probably relate to this phenomenon.

Finally comes the authorization for the start of production. Let's say the total sales of the product are expected to be $10 million. Someone with the authority to authorize $10 million should sign the final production-release form. Does this not sound like a logical, commonsense requirement? Otherwise a product could be "nursed" into production in dribs and drabs, with the company eventually involved in a huge business that no one has ever really approved as a whole. We'll talk about what constitutes a "production release" in the "Configuration Management Systems" chapter.

In summary, any procedure for product development should be divided into a small number of logical, phases, each representing a significant block of expense. The reviewer of each phase needs to be able to authorize the spending for the next phase. Otherwise, why review that phase at all? I hope that you'll see that this is a common-sense approach to product development procedures.

Follies and Foibles of Design Reviews

As with some other subjects in this book, design reviews are often very painful for the practitioner and equally useless for the organization. Why is that? A design review, from its very name, would seem to be a very logical and useful thing to perform. Certainly during the process of developing a new product—be it a mechanical gizmo, skyscraper, or software package—it would be prudent to review the features of the product and judge whether they meet the requirements. I maintain that the problem lies in the difference between the perception of management and reviewers and that of the "reviewee." It is all too easy to turn a "design" review into a "project" or "program" review; and therein lies the rub.

A design review is a static process. It is a review of a usually dynamic thing, a product, or a process at a snapshot in time—presumably right now, at this moment in time. The appropriate question is, "Does this dimension, that feature, or this line of code meet the given requirements?" Questions never to be asked are, "Have you done this, tested that, or completed another?" Those are dynamic questions, not to be allowed in a true design review. In contrast, a "program" review is totally dynamic in nature. It is all about progress against a predicted timetable, against a budget, or in a change in circumstances. The difference in preparation and presentation is dramatic. The problem, though, is that it is very easy to merge the two or drift between the two, even in the midst of what started out to be a design review.

The one dynamic feature that belongs in a design review is the specification itself. I have yet to see a customer requirement that was complete at the beginning and unchanging (static) for the life of the project. Remember, a specification is essentially a piece of paper that attempts to describe all the attributes of, say, a paperclip, computer program, or skyscraper. It's simply not reasonable to expect that the requirements won't be refined or perhaps completely changed in the midst of things. So in the instant in which the design review takes place, the two dynamic things— the requirements and the design—are frozen and compared with each other. The past doesn't matter, and the future isn't yet here. If the requirements have changed so much that the design no longer applies, this is the time to point that out. The design can be rejected completely, or a list of inconsistencies might be generated. The timetable for the reconciliation to be done is in a *project* review—not the design review.

The above more precise and narrow definition of the design review may help reduce the personal stress involved and increase the value of the output. However, there is one more element in the design review: the audience. I'm sure the presenter would prefer that a single person be present—his or her direct management, in which case both the expectations and the results would be seen more clearly. Alas, that is seldom the case. I have presided over a great many pure design reviews with my direct reports, and I always felt the reviews were very productive (I wonder what my subordinates thought). Usually, though, the audience is a "cross-functional team of peers," a phrase sure to generate dread in the mind of the reviewee. And often the "team" is expected to come to a "consensus." That's a fine kettle of fish, and I can think of three serious shortcomings of this scenario (but give me a minute and I'll think of more).

First, all members of the audience will never have the same level of expertise as the reviewer, and that means the usefulness of having them review and pass judgment on the design is questionable.

Second, the fact that they are "peers" means that to some degree they are competitors, bringing into question their motives in any active involvement. After all, if your competitor—I mean peer— looks too good, he or she will get that promotion you've been angling for.

Third and finally, I have to question the whole concept of "consensus." By definition that means that all present must agree. Anything less than total agreement, begrudging or not, can stop the design review in its tracks; 100 percent agreement means the design is OK, and anything less means the design is not OK—unacceptable. Then what? One problem with the consensus approach is what I call the "broomstick in the spokes" phenomenon. It is like the reviewer is riding a bicycle down the street with the audience sitting along the curb, each with a broomstick in their hand. Anyone can, at any time stick his broomstick in the spokes and stop the whole thing. And it's tempting to do: a peer, who normally has no control over you or your success, is now empowered to stop you in your tracks. Very tempting.

With that I have to recant and backtrack just a little. Having a cross-functional team present can serve a purpose. For instance, the purchasing director might want to state that the supply of unobtanium is getting scarce and the design should use some other material. Bringing outside perspectives into the review can be very useful. In theory (my theory, at least) a design review doesn't result in approval or disapproval. It merely results in a conclusion about how well the design meets the specifications. After all, how can your peers approve your work? That's a job for your management.

I have been present when almost all elements of a design were short of acceptable, but the design review went to completion and everyone called it a "good review." Well, maybe the mechanics of the review process went OK, but that's *not* what was supposed to be graded. It's the design itself, which can be acceptable, unacceptable, or acceptable with required changes.

Yet another problem—I told you I was going to come up with at least one more—is that no one wants to be confronted and asked for a decision in public. After going through a number of reviews the hard way, I finally arrived at a design-review strategy that seemed to work. I took the time to do a "pre" review one-on-one with each person who would be in the formal review. As concerns came up, I made sure they were included in the review, whether they were resolved or not. Almost all the time, the formal design review went off without a hitch. Everyone was impressed and felt a part of the process, not just an outsider looking in. Most importantly, no one was asked to pass judgment in public without having a chance to study the subject.

There you go—design reviews can be useful and helpful without being painful. The trick is to stick to the objective and manage the human factors effectively. Easy? Maybe not, but at least it's not all that painful. As for the inherent interference between the objectives of a design review and a program review, take a look at the chapter "Product Development Procedures."

Specifications: What Do They Specify?

The type of specification most people might think of first is a dimension of a part, so in the following discussion, I'll use that as an example. However, specifications can be any description of a requirement. Specifications are usually accompanied by limits—after all, without limits, specifications would have little meaning. One might think that the specification "must be free of visible defects" is a good one, but it has no limits. A better one might be "must have fewer than 10 visible blemishes, none larger than .010." And finally, to enforce the spec limits, there must be some penalty for exceeding the limit. So let's compare a couple of spec limits and their common usage.

A part is usually interpreted as being "good" if its measured characteristic is within the spec limits of the required dimension and "not good," or scrap, if it measures outside a spec limit. Hmm, let's think about that for a second. Two parts could have a characteristic that measures infinitesimally different, and one will be used and the other thrown away. Logically, that makes no sense; but in fact, that is the common usage. Most part populations will have characteristics clustered about their mean—after all, that's our objective, right? So that means the parts out of specification will be at the "tails" of the distribution. If you accept that the parts barely outside the specification are probably acceptable (after all, the parts just inside the specification are acceptable), then *most* of the parts rejected will be good parts! Discouraging, isn't it?

Let's pick another specification: a project has to be completed by a certain date. That's a one-sided spec as it has only a max spec limit, not a minimum—perfectly legitimate. So let's say the contractor finishes one day late. What is the ramification? No payment? But the project is only a little different than if it had been finished the day before. Generally, large construction projects have a late-completion penalty, which means the contractor still gets paid, just a little less.

So why would you throw away the part that measured just a little outside the spec? Conversely, why would you keep the part that measures just a little inside the spec? There is no good answer, and therein lies one of the frustrations of specifications.

It must be recognized that specifications are at best a poor and approximate representation of the actual requirements. Why did the construction project "have" to be done on that exact date and not a single day later? Why is the tolerance on the dimension 0.01 and not 0.01001? The idea, of course, is that the part (or project) won't work if the characteristic is out of specification. In the case of a dimension, that might be true if, say, one part is past the upper limit and the mating part is past the lower limit of its corresponding dimension, making the parts interfere and preventing assembly. But what are the odds of that? Any discussion of specifications then becomes dependent on statistics.

Then there is the question of what the specification means in the first place. For example, it is all too easy to place a diameter specification on a nominally circular part. I once asked an inspector how he measured a round part to evaluate it against a diameter specification (true story). The inspector replied, "Easy. I just measure across the part in three different places and report the average. If the average is inside the spec limits, the part is good." Really? The inspector could have measured an equilateral triangle, or more correctly a Reuleaux triangle, and called it good—after all, if he measured at any position from one side to the opposite it would always measure the same.

To add to the story, the part was molded in plastic with dies that opened transversely, creating the inevitable parting line on the "diameter." The inspector added, "Of course, I don't measure at the parting line, as that would give me an error." That would make most people speechless—it did me. In fact, evaluating a "circular" part against the typical diameter-with-tolerance specification is extremely difficult, and I frankly don't know the absolutely correct method. The truth is that round parts have historically been made on a lathe, making them inherently round and concentric; over the years we have become lazy with our specifications. Maybe we should look for a way to specify a part that is less ambiguous.

Readers concerned with nonmechanical parts only can skip the following discussion, but it is a good example of how a nontraditional method can be effective.

In this case, the subject part was a "cylindrical" probe (*cylindrical* is in quotes since nothing is truly cylindrical) that incorporated a groove for an O-ring. The part was molded from plastic, which didn't really change the problem but did make the proper specification methodology more critical. The main criterion was to keep the O-ring from leaking. Just to make this discussion manageable, let's assume the mating port was circular and had upper and lower diameter limits.

For the O-ring to seal, it had to be squeezed at least a minimal amount (don't use the squeeze dimensions you'll find in O-ring handbooks—but that's another discussion) but never so much that the rubber would "fail" in compression, causing it to take a set. Also, note that one part of the O-ring had no idea what was happening at some other part—each and every location on the O-ring had to be squeezed appropriately to make a seal. And the probe could move in the port so it was against one side. Naturally, if we measured from the port outer surface to the groove on the opposite side, the dimension had to be at least the maximum port diameter, minus the O-ring section, plus the minimum required squeeze. Note that the specification was single-ended—it only specified the minimum limit, and it was a "constant" dimension, meaning that it applied to all orientations. So far, so good.

Next we needed to make sure that the O-ring didn't get squeezed too much. That meant the depth of the groove at any location had to be at least the dimension that produced the maximum allowable squeeze—another single-ended spec. Remember, the probe couldn't be assumed to be centered in the bore, but it could be pushed to one side. That was the second requirement—pretty simple, right?

The third requirement was that the probe had to fit in the bore, so the distance measured across the port at any location had to be less than the minimum diameter of the port—a third single-ended spec. There you go: three dimensions, all single-ended and each equally easy to measure, which completely described the requirement of locating the probe in the port and preventing the O-ring from leaking. Think about it—how could you even build a part that meets those three simple specs yet doesn't perform the function?

Now let's look at a more conventional specification methodology. First you would likely provide diameter and roundness specs for both the probe outer surface and the groove, a total of six dimensions (four specs, each with an upper and lower limit, plus two single-ended specs). Then you would add the concentricity requirement to center the groove in the probe, making a total of seven specifications. To keep the O-ring from leaking, the tolerances on each characteristic would have to be quite small, or the stack-up allowance would be unacceptable. Now the molder of the plastic part would have to create a tool that would meet all those requirements.

Compare the difficulty of meeting those seven dimensions with the simplicity of meeting the three easily measured dimensions previously described. This actually occurred, and the biggest difficulty encountered was convincing the draftsman to actually put the three nontraditional dimensions on the drawing. Finally, we compromised and left the conventional dimensions on the drawing as a reference. The molder had absolutely no problem building the part, and after several million parts were built, none of them ever had a leak.

Consider another example related to the meaning of typical specifications. This one also involves a dimension on a mechanical part. The two sides of a part appear to be parallel on the drawing, and a dimension shows the width. This should be simple, right? Just measure across the part and compare it with the dimension. But wait—the sides might not be exactly parallel, so what then? Look for a parallelism requirement, right? This is already starting to get complicated, so we measure the width at each end, which has to be within the width tolerance limits, and the difference has to be within the parallelism limit. But what happens if there is a bow in the surface? You might say we need to add a flatness requirement. Wait, let's back this pony up a bit. All we really wanted was a thickness. Why not simply require that the thickness be met at every point on the part? Done.

Below are two sets of specifications. The upper picture shows a conventional geometric dimensioning and tolerancing (GD&T) methodology. Looks pretty simple, right? But in the boxes are actually additional requirements, each of which has to be proven capable by measurement. The lower picture shows it as it ought to be...OK, as I say it ought to be. Three simple, easy-to make measurements can be used to prove capability. But how do you report a "constant" measurement? Use the average? No, no, no. Use the worst measurement of all the ones taken. Remember, one measurement out of spec will make the part not work, so it's the worst that counts. For instance, on the ".500 max constant" specification, the highest dimension measure would be reported, *including* taken across any molding flash.

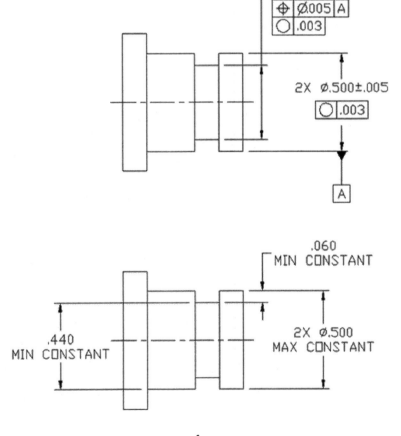

But are we really done? If we were to put a requirement on the specification, we would need to be able to agree with our supplier and customer on how to verify that the requirement is met. On the thickness requirement above, note that any material will expand as the temperature changes. Does the thickness have to be in spec regardless of temperature? Say the surface could wear over time—does the spec have to be met over the entire life of the part? The answer would be yes, unless you specify the condition under which the requirement is applicable.

Another difficulty that can be encountered is when a "top drawing," or top-level specification, is used as the interface between the customer and supplier. The parties have their own expectations for the document, and often they are in opposition.

A good example might be the voltage requirement for an electrical part. The customer could (and often does) say something like "4.5 to 5.5 volts" is the voltage supplied to the part. Does that mean that the part must meet all the requirements at any voltage between 4.5 and 5.5? Unless there is language to the contrary, yes, it does. So does the part have to function at any voltage up to 5.5? That sounds like an upper spec limit, but actually the part has to function at a voltage of *at least* 5.5, and that sounds like a lower spec limit. "What's the difference?" you say. "We all know what was meant." The difference arises when we qualify the part with a power supply that has its own spec of, say .1-volt maximum error. To guarantee the part functioned at 5.5 volts, the test system power supply has to be set at 5.6 volts, and then it could be supplying 5.7 volts. So if I were making and qualifying the part, I had better design it to be functional at 5.7 volts. Just another example of how things can get complicated.

The point of all this is to illustrate that even simple specifications can become very complex. It is always a struggle to simplify them and yet keep them totally descriptive of what you expect as a user. Many examples exist where someone writes a list of requirements in a few minutes or hours and considers that task done; then it takes weeks of struggle to remove all the ambiguity. But often at the end of the specification review, the specifications become much less complex than they were originally, and that's the way it should be. Remember, before the product is built, the specification is only paper. What seems to be an agonizing process of fine-tuning each minute detail in the spec will almost certainly pay off later. So I beg of you-agonize over every detail of the spec.

FMEA: Trials and Tribulations

Creating and executing a failure mode and effects analysis (FMEA) can be the most frustrating, demanding, and time-consuming task in a project and yet result in—simultaneously—the most useful and useless of outcomes. How can that be? We're going to attempt to find out in this chapter.

First, what could possibly be wrong with analyzing the modes and effects of all possible failures? And what can be more integral to any project than keeping in mind the possible ways things can go wrong? It is almost a certainty that the first time a designer thought of what could go wrong was within seconds of the conception of the product, so the goals of an FMEA are integral to any development process. The following discussion is centered on the premise that something must have led such a simple and fundamental concept astray from its original purpose. Perhaps it has become overorganized and regimented to the point that the original objective has been lost along the way.

Let's take a look at some of the elements of the analysis. Certainly, the fundamental concept makes perfect sense—rate the potential failure modes by severity, likelihood of occurrence, and probability of being detected and the consequences avoided. Multiply the three factors together to create what is usually labeled the risk priority number (RPN), and then rank the failure modes so that those with the highest priority (high RPN) get the most attention. This is a straightforward concept and one that can't really be argued against. Right?

First, what is a failure "mode?" It might be best to think in terms of what it's not: it's not the result, and it's not the cause. So the part not "working" isn't a mode—it's the result. Selection of a material insufficient for the task isn't a mode—it's the cause. "Breakage of the material" could be a mode, but a little more definition is in order. "Breakage of the lever in bending" is a more clearly defined mode. Sounds straightforward, but identifying and clearly defining each mode is not always easy.

The severity of a failure mode is the first characteristic to rate. Using the above example of the lever, does the failure cause a slight inconvenience, or does it cause potential injury or death? Or something in between? Herein is one of the weaknesses of most FMEA analyses. Often, severity ratings go from one to eight or so. You could use zero, implying that nothing whatsoever happens when the lever breaks, but that's not likely. Generally, a "one" rating is when the failure is barely detected by an expert—and by the user not at all. A "seven" might be when the device abruptly stops working, leaving the customer stranded and greatly inconvenienced. An "eight" is then reserved for something that can result in bodily injury or death. Remember that the severity number gets multiplied by the other factors to arrive at the RPN for that failure mode. So let's get this straight: according to most ranking systems, if you are the user, death is only 14 percent "worse" than being highly inconvenienced! How can that be? Many published FMEA systems add corrections to overcome weaknesses like this one, such as requiring every mode with an "eight" severity to be placed at the top of the list. That doesn't make sense to me—the FMEA system should systemically do that without adding exceptions and special cases. Instead, I recommend the employment of a geometric rating system, usually starting with one. Then comes two, four, eight, and sixteen. In that way, the more severe failure modes are dramatically highlighted, as they should be. You might even decide that each incremental severity be higher by a larger factor—or maybe by an exponent. Just make sure that higher severity numbers are somewhat proportional to your perception of the actual severity.

Next is the likelihood of failure, or frequency. Most agree that the rating number should be proportional to the frequency, but here again there is a tendency to assign a linear progression of numbers to failure rates that might go up geometrically, by a factor of perhaps ten. Say "four" is a rating for 1 failure per 10,000 while "five" is the rating for 1 in 1,000. Why not just put the frequency down as a rating? Thus, 1 per 1,000 becomes a rating of .001, and 1 in 10,000 would be .0001. Or to make the math easier, just put down the failure rate in parts per million. That keeps the rating directly proportional to the magnitude of the problem. One thing to look aggressively for is a way to eliminate the part that exhibits the failure—in other words, keep adding more and more simplicity. Then the frequency could truly be changed to a rating of "zero."

And finally comes the detection rating, which is a little bit more subjective. A low or zero detection rating means the failure is detected *before* the customer can find it, or perhaps even before the next manufacturing process. Will accurate stress analysis detect the potential failure mode of the breaking lever? If so, the failure mode might be rated a "one" or even a "zero." If stress analysis is not considered adequate to absolutely detect it, then a higher number would be used. And here is where it gets subjective. Exactly how likely is it that the failure mode will be detected? And notice we are talking about detecting, not solving or preventing—that's the object of later evaluation. It's very tempting to use the detection number to "adjust" the FMEA results to match our preconceived sense of reality. Say the detection rating is "two," but the RPN ranking puts this failure at the top, even though we are all confident it isn't a real problem. So, if we just increase our confidence that it can be detected and change the rating to a "one," the resulting RPN will be half what it was before. Tempting, but is it realistic? If you use the FMEA process to simply reflect what you already believe, you are wasting your time and not taking advantage of the power of a formal FMEA.

But how do you get started doing this usually agonizing process? This is where the rigor of the formal process departs from a logical, practical, and effective (commonsense) route. Usually, the first time someone thinks of a potential failure is about a second or two after the product concept comes to mind. Thinking about potential problems (failures) is integral in the development process and certainly something we all do. So why not write it down (OK, enter it in the FMEA form or make up your own spreadsheet) right then? The failure mode isn't well defined? Document it anyway, in however crude the form, to be refined later. Don't have a good idea of what to put down for some of the rating numbers? Don't worry about it; enter numbers that represent your best guess at the time or leave them blank, to be entered later. The point is that the FMEA as a document should be progressively created as an integral part of the development process. It is a repository of what is learned about potential failures and their prevention. Naturally, every time a failure actually occurs during the development process, it should be immediately entered. *This is key to the effective use of the FMEA process.*

What happens if a development occurs that changes the nature of a failure mode or the rating numbers (hopefully in the downward direction!)? This gets a little tricky, depending on the configuration of the software, if any, that you are using. I believe it is important to keep the existing information and not just overwrite the data—which is not too difficult until you start to manipulate the spreadsheet, such as reordering the line items by RPNs, which is a natural process. Some people create another spreadsheet so that "old" line items can be moved there. How you do it is up to you, but you will want to keep a permanent record of all the entries and modifications. Formal FMEA tools might do this better than home-brew versions.

Bottom line: if you don't get wrapped up in the formality of the methodology, and if you start the FMEA at the earliest possible time, it can be a highly useful development tool. This is regardless of whether your product is a car, software package, or fast-food service. It is much better to have an FMEA handwritten on scraps of paper that is actually used to guide the development than one in the most formal hundred-page document that covers a myriad of possibilities but is never again looked at. All to often a product is launched with an elaborate FMEA filled out after the project was complete—and just as often the very first field failure wasn't even anticipated by the FMEA, let alone given a high ranking. Now *that* can be discouraging.

Once you have an FMEA created, what are you going to do with it? Often, what happens next is exactly nothing. The FMEA is created after the completion of the program in response to a requirement from an apparently vindictive and sadistic customer who merely wants to inflict pain on the supplier. It is then paraded in front of the customer to show that there are no high RPNs, so "What could possibly go wrong?"

One can study without learning, learn without understanding, and understand without acting. In the end, action is the only thing that counts. The purpose of the FMEA is to identify the highest risk factors so that corrective actions can be prioritized. The corrective action and result is then added to the FMEA, usually resulting in a reduction of one or more ratings—the essence of a useful FMEA. It is a rigorous method of assigning priorities and tracking the progress of a project as it relates to potential failures and their solutions.

Someone is sure to say, "Well, that's OK for a hardware-development project, like for a car, but I'm doing this software package that doesn't really 'fail' or cause injury or death." This is simply not true. What if the computer program is intended to be faster than any other—for instance, execute an order on the stock market in one millisecond instead of the competition's two milliseconds? A glitch in the code that makes a subroutine hesitate for three milliseconds is sure death for the program and likely deserves the highest severity rating.

An anecdote (to be read at your own peril and amusement, or simply ignored): During the development of an engine-control system that operated the ignition system, the code writer wrote two subroutines—one that was used at low engine speed and one at the higher speeds. Both worked very well, but his supervisor (that would be me) asked how it made the transition. After a bit of fumbling, the "expert" said it could, if the timing were just "wrong," skip an ignition event—"But don't worry. It will pick it up on the next engine revolution." To him it was a minor and inconsequential failure, but to me it was catastrophic since the customer would never tolerate even one misfire—however unlikely it might occur. The code writer would have rated the severity a "one," while his supervisor rated it a "sixteen," thinking it was closer to "severe injury or death."

So who gets to decide on the severity, occurrence, and detection numbers? There is often a difference between theory and practice. In theory, customers should determine the severity numbers because only they know the effect of the failure on their system and the final customer. But left alone, the customer will likely err on the side of higher severity—after all, the customer doesn't want to be later criticized for saying any failure is not "severe." So the severity numbers might best be arrived at by consensus. The product designer would be the logical one to select the occurrence number because he or she knows the intricacies of the design and the likelihood that an event could occur. Of course, the problem is that the designer, also being a normal human, is likely to rate the occurrences very low (again, "what could possibly go wrong?"). In the end, all the rankings have to be collaboratively decided by the team that includes the designer, manufacturer, quality department representative, and customer.

The less agony that goes into a FMEA, the more effective it will become. I guarantee it.

Configuration Management Systems

Companies, or even activities, require some form of documentation system, written or not. There has to be some method of creating consistency, commonality, and control that transcends individual ideas, methodologies, and memories—and even individuals themselves. In many ways a company is similar to the US government. The United States is not a monarchy and in fact not even a democracy. Rather it is a republic—a country governed by law (the constitution). Just as the government does, the company must continue on a true track if someone quits, gets fired, or goes away for some other reason.

So, what should this documentation system be like? Just as someone has to be at the "top" in every organization, some document must be at the top that sets the position of all other documents in the hierarchy. In some organizations, this document is called the "quality manual," but it doesn't have anything particularly to do with quality. It might better be called the "constitution" of the company. Whatever it's called, its purpose should be to set the standard for other documents. The most important characteristic of any document is the level of approval required for it to be changed. For the top document (TD), it is only logical that it be approved only at the very top of the organization. The owner, president, board of directors, or whomever is at the top should be the only ones who should approve this document or changes to it, though, like any other document, it can be written or proposed by anyone at any level in the organization.

So what is in this document? It might logically start with a list of company goals and objectives. Next might be a list of departments and the charter for each. But the nitty-gritty is a list of document types and the signature authority for each. Some types include procedures unique to a given department, and the head of that department would be the ultimate approver. Some, such as the quality-control policy, might be approved by the heads of all departments. The important feature of this TD is that, for instance, the head of a department does not get to decide which documents he or she approves—it is preestablished in this "constitutional" document. And most certainly the writer of a document does not get to decide who will approve it.

Some function has to be set up as the "keeper of documents." Quite often, the engineering department is tasked with the document-control process, probably because engineering documents usually constitute the bulk of the workload. However, I suggest that a separate department be set up as the official "library" of the company. In a small company, this might be just another task for an administrative assistant, while in a big company, this could be a large, centralized activity, even independent of each operating division. Regardless, it is a function that should be immune from individual influence. Like a city library, it doesn't write the books; it just keeps them in a safe place. While this function is often labeled "document control," a more descriptive term might be "configuration management" as it documents and controls the configuration of the company's processes and products. Its purpose is to keep documents "locked up" in a safe place and to maintain the integrity of each document.

How should they be kept locked up? There must be a procedure described in the TD that defines who can approve what, and this is where the controversy starts. Some will argue that the document itself should list the required approval signatures, but that would seem like putting the fox in charge of the henhouse. Others will also argue that there is no need to involve functions not affected by the document. OK, but who is best capable of that determination—the person or department in question, or the writer? I suggest that all department heads be required to approve most procedural documents, as only they can determine if a procedure affects them.

The method used to identify or label each document is probably less important in this era of electronic storage than it has been in the past, but there is much wasted effort, usually resulting only in confusion, expended in creating numbering systems. It is tempting to create a system wherein the label or number has significance. "It is easier if, when I'm looking for a quality-related procedure, I look under the Qs." What? Just do a search for documents with, say, "quality inspection procedure" in the text—it's much easier and faster. Why not just number the documents sequentially and get it over with? Of course nothing is absolute, so you could use a prefix of C for corporate-wide documents and D for departmental documents—or something like that.

Many companies are involved in the manufacturing or sale of physical components and assemblies, and these companies need a way to number the documents that define these products and components. The most logical way is a simple sequential numbering system that includes both components and assemblies. Furthermore, the number should be assigned by the document-control group when the document is released, not when it is "checked out" by the designer or engineer. When the latter is allowed, designers often check out, or assign, a whole group of numbers to an assembly they plan to create. The argument is that it is more logical to make an assembly drawing with known numbers listed in the drawing and then "release" all the drawings into the system at once. And the components can have sequential part numbers, which somehow satisfies the designer's sense of order. OK, but what if things change, and some of the numbers don't get used? That creates blanks in the numbering system, which brings discontinuity into what should be a continuous sequential system.

Another note about part drawings or specifications: they should never "point" up, only down. For example, an assembly drawing can identify a component part number that is contained in the assembly, but a component drawing should never contain the assembly number in which it is used. If the assembly stops using the part or the part is used in some other assembly, it would become necessary to revise the part drawing even though nothing about the part has changed. The same is true of customer or approved-vendor references. Just keep it simple—the part drawing describes the part, nothing else. I suggest creating another controlled document that lists approved suppliers for parts. That document will likely be changed often, as suppliers come and go, but the part drawing won't need to be changed.

And my golden rule is that you should be able to pick up a part (look at the software code, etc.) and determine whether it meets the drawing's requirement. You should never put in process requirements like "select fit" or "100 percent inspection of this feature." Those belong in a control plan or process description. In other words, if the fit is so precise it requires a select fit, the fit itself should be specified, not the process by which it is accomplished.

What in the world is a document "release," and where did that word come from? Who knows? But it must be something like releasing an animal into the wild, except that in terms of documents, it is when the document is "captured" and locked into the document-control system. The real questions are, what is a controlled document, how is it created, and what does it mean?

In principle there are only two locations for documents (whether they are drawings of parts, specifications, software code, or whatever): either inside the "vault" or outside. Any document worth preserving usually progresses through multiple levels of "control."

To begin with, it only exists in the mind of the creator (no, not that Creator) and has no control at all. The person's mind can change in a microsecond, and the document will change with it—no outside approvals required! Then the creator decides to write it down—or these days commit it to electronic ones and zeros. While working on it, the creator will likely find a place to keep it, which therefore imparts some degree of control. Whether the individual creates a rigorous labeling and revision system as he or she goes along, or just attaches a label to it, it is the creator who has complete control. Often CAD designers will create an elaborate system of keeping track of their drawings. So far, no problem.

Sooner or later, though, the creator will find a need to let the document "out" into the system so others can use it. How does that happen, and who is then in control? Again, that's where the controversy starts. In my opinion, the designer should be permitted to simply take it to the document-control person, group, or department and say, "Here it is; put it in your system (the vault)." Some might say, "No, it has to be approved." Why? So far, nothing has been done, and no one is required to do anything as a result of simply storing the document in the vault. So why does it have to conform to an approval procedure? Just record who put it in the system and when. By putting it in the system, the document is transformed from an uncontrolled to a controlled document. And that means that from that point on, changes must be documented and approval of those changes must follow a procedure (refer to the previous discussion on procedures). This procedure should be contained in the TD, since it can affect all departments.

Control is not absolute—there are degrees of control. The first level of control is often called "experimental" or "preliminary." This implies that changes are expected to be made or that the requirements in the document are not yet to be followed except at the user's risk. But the first level of control is very important for a couple of reasons that continue on through all levels. First, the labeling system of controlled documents is company-wide and is both specific and nonredundant. Only *one* document per label is thus identified, and that document is unambiguous. When someone asks for document 1234, they will get only one, and everyone in the company will know what they are talking about when it is referenced. Second, changes are tracked, so anyone can talk about rev A or rev B, and people will be able to find the document and know what is being talked about (nobody has to physically give them the document—they can get a copy for themselves). How do changes to these "first-level" documents get put into the system? Allowing only the creator to change the document can create problems, either with a proliferation of changes or in the event the creator leaves. Some think this can solved by having a "one up" approval required—someone one level higher in the organization must approve the change. I suppose that is a reasonable approach, although as you can see, I'm not all that enthused about adding that level of restriction.

The next level of control is often the only other level and is usually called "production released," or something like that. This label is not good since it implies that the control level has something to do with the authorization for production. Perhaps you could insert a third intermediate level of control, but the justification for such a move might be difficult. So, let's say there are only two levels of control—level 1, where changes can be approved at a lower level, and level 2, where changes require a higher level of approval. And moving from level 1 to level 2 requires the same approval as changing an existing level 2 document. OK, that sounds logical, but what kind of "authority" does the move to level 2 give to the document? It probably should depend on the type of document, but essentially, when a document gains level 2 status, the whole company agrees to comply with the requirements in it. If it is a procedure, all groups are required to abide by it. If it is a part or product specification, all groups are required to recognize it as such ("Yes, that's part 1234, and that's the only part 1234").

But here is the rub: in many manufacturing or production organizations, when a part is "production released," there is tacit, or implied, approval for it to actually be produced and sold. This is *wrong, wrong, wrong!* The release merely says that from now on this part/procedure/specification will not change without approval from all groups (OK, specified groups) in the company. Remember, a document-control system is a "static" system. It can only say what *is*—it cannot say what *to do*—and it cannot give instructions or approval to anyone to actually do something. In other words, someone can create a document that describes a part or gives instruction on how to do something, but that doesn't give a command to anyone to go out and actually do it. In the case of procedures, the TD or company management will say something like, "We will abide by and follow approved procedures," so it is implied, but not specified in the document, that people will actually perform the procedure. It does say that *if* anyone is to perform the procedure they *must* follow the documented process. In the case of a production part or product, the creation of a document that describes the product does not by itself give the command or even permission to build and sell it.

This is the principal difference between common usage and what should be common sense.

Let's look at an example of how not to do it: In a company, part 1234 is destined for production and sale, and procedures say a "production released" document approves the part for production. Before that happens, tooling and capital equipment must be purchased at great expense. Then the part built from that tooling must be put through—and pass—a rigorous test protocol before it is allowed to be sold to a customer (often the customer will demand this). But per company procedures, the part drawing has not yet been "released" in the document-control system—how could it, since the qualification testing must be completed before it can be produced for sale? This is a conundrum: the drawing must be released to guarantee conformance to the design, but per company procedures, it can't be released until approval for production. Hmm...this means there are drawings being used for building expensive tooling and subsequently for product-verification testing that can be easily revised at a low organizational level. Sure, the responsible engineer may be intimately familiar with a change and convinced it is justified, but he or she didn't write the check for the tooling and might not know the implications of any possible delay in the program. So he or she could easily and innocently make a change that could increase the cost of the program and cause significant delay. Something is seriously wrong here.

So here is the solution (OK, *my* solution): Create two documents, one that describes the product and another that controls the implementation. The specification, product drawings, and process documents are *static* ones, and the others are *command* documents. The specification says, "If you want to talk about part 1234, this is what it is." The production release says, "Build part 1234, go forth and conquer, I command of you." It is an action, or command, document. The specification neither knows nor says anything about schedule, pricing, or cost. The command document is *all* about cost, pricing, and schedule.

Back to the example of how not to do it: Once in production, sometimes the product dies a graceful death; then years later a production order for it appears in the system. What happens next? The sales group could logically assume that every "production released" part is available for sale and can be manufactured, so they create a price, quote a new customer, and receive the order. What kind of control is that?

Here's the way it should work: First, the product is defined and approved as a level 2 controlled document. Then upper management decides it is worth the gamble, based on business considerations and whatever other factors they deem appropriate, to spend money for tooling that will be completed in time to meet a real or perceived customer need. Then they review all considerations—for example, whether the part passed qualification testing, the customer still wants it, and the pricing is still viable (will it make a profit?)—and then, and only then, they sign a "production release" document that says, "Yes, go forth and conquer." At that point, the entire organization springs into action and buys raw material, builds the part, and sells it at the price specified in the production-release document. Then suppose another customer wants the part. In the old system, the sales department would likely just give them the price for which it has been selling and go on. But in the "new world," the production-release document is *not valid* for the new customer, since it specified, among other things, the customer, price, and delivery. So a change order has to be written to modify the production release to include the new customer and probably new pricing, delivery, and so forth.

In review, a document-control system must be both unique and unambiguous—that is, there must be only one of anything and that thing must be completely defined (there is only one part 1234 and the documentation describes only part 1234). The document system is a *static* system—it merely (merely?—it's pretty important) describes a part, product, or procedure and does not allow or command anyone to do anything. The command to implement is a separate function—but a function that is still embodied in the singular configuration management system.

There—that's it, a commonsense approach to configuration management. Simple, eh?

Single-Piece Flow in Manufacturing

What is this subject doing in a book on quality? Perhaps just to illustrate diverse ways of approaching a task. Unless you are involved in a manufacturing process you could probably skip this chapter with little ill effect.

Whenever any significant number of product is to be manufactured, the argument of single-piece versus batch processing seems to crop up. While it could be true in any situation, in a manufacturing environment, *the easiest decision to make is usually the most difficult to implement*—an obvious corollary to Murphy's law.

Implementation of batch processing is often the result of rushing to a simple decision. The reason is often that in the experimental and prototype phase, only a few parts are processed at once, usually in small batches, so it might seem logical to just increase the number in a batch—no reason to clutter the mind with worries about process-time variability, machine reliability, throughput mismatch, and so on. Just put multiple components in a tray or fixture, process the parts, and take them to the next process. But what kind of fixture? The same fixture for all processes? Sounds good, but one fixture usually won't work for all processes, so more than one type of fixture will likely be required. Manual loading? The question is often not allowed at this point in the process—it's already making the simple decision complicated. I hope you can see how making the simple decision up front could make the subsequent implementation exceedingly complicated and difficult.

Conversely, in some cases, batch processing can be justified. Perhaps the processes are inherently separated by a long distance, for instance the coal mine's distance from the power plant—it's easier to put batches of coal on a train than to build a thousand-mile conveyor belt. But choosing batch processing can lead to large problems later. What kind of transport medium will be used? Some kind of tray, box, truck, or train? Is the easiest method of putting the parts in the fixture also the easiest way to take them out? Can the tray be used in the process itself as a fixture? How far along in the process can the fixture be used, or will a different fixture need to be designed for the next process? How much will these fixtures cost, and how many will be required? Will all machines be able to handle the same batch sizes? How much in-process inventory will be required? If a part is rejected, will the fixture now have an empty space, or will the rejected part continue, just to take up the space? How will the parts be protected while being stored, waiting for the next batch? Will they degrade over time? Suddenly, one-piece flow doesn't seem so complicated.

The design for a single-piece flow process isn't simple either, but the difficulties are readily visible at the outset. Rarely, if ever, will two processes have the same throughput. All machines will exhibit a different degree of process-time variation, reliability, and durability. Do subsequent processes need to be synchronous? A synchronous process flow is one in which all processes are performed at the same time, all parts marching through the series of processes in lockstep. It's easy to visualize but fraught with problems—remember, "The easy decision is often the most difficult to implement." Obviously, the speed of each process is limited to the speed of the slowest process. If any process has any variation in process time, all processes have to accommodate.

A number of books have been published that discuss process-time variability of series processes. If each process has random process-time variation, there are a few ways to accommodate it. One way is to figure out the longest time any process could take and slow the line to that speed. Another is to pick a line speed that is "reasonable" and then simply throw away a part if it requires more than the allocated time. A third is to wait for all processes to be complete and then move all parts to the next step. About this time you should be thinking, "Are you nuts? None of those options are acceptable." And they shouldn't be—although I've seen processes run in all three of these ways.

But don't despair; there is another way. And that is to implement single-piece *non*synchronous processing. This requires a queue or buffer ahead of each process. Right now you might be thinking, "Wait, to add dozens, hundreds, or thousands of parts in buffers between processes will negate any benefit of one-piece flow." If so, you would be wrong. Here is a logical way to implement the process queuing: Each machine accepts a part as it arrives and performs its process without regard to anything that happens anywhere else on the line. If there is room in the downstream buffer to accept the part, it sends the part on its way—if not, it pauses, holding it until there is room. And if there is no part in the upstream buffer, it waits until there is one and then proceeds to process it. The result is that each process is completely independent and performs its process without regard to what happens elsewhere.

Now *that* is true simplification. The ramifications are enormous. If one machine has a stoppage, the buffer in front builds up, and one by one the machines come to a stop, either because they have no part to process or no place to put their processed part. But if the fix is quick, the machine can be put back online without any of the other machines even knowing that it stopped. The slowest machine will tend to collect an upstream buffer and will run continuously, and the line will then run at the average speed of that machine—what could be better than that? To increase the throughput of the line, another identical "slowest machine" could be put in parallel, with the upstream buffer then feeding both machines. Then there will likely be a different slowest machine. Consequently, line-speed improvements could be ongoing, causing minimal disruption. Such a system would be simultaneously simple, effective, and elegant.

So, tasked with configuring a production system or similar process be alert when making that first decision – the easiest decision to make is often the most difficult to execute. Take care and use your common sense – it will pay off in the end!

Efficiency versus Effectiveness

The terms *efficiency* and *effectiveness* are often used loosely and interchangeably. But should they be? What's the difference between them?

Efficiency, by definition, is the quantity of effective work output divided by the input. It's usually relatively easy to understand. In the case of an engine, efficiency equals the work done by the crankshaft divided by the amount of fuel burned. This measure is often expressed as "brake-specific fuel consumption," usually brake horsepower divided by pounds of fuel per hour. In the case of a car, it might be expressed in miles per gallon—after all, the "useful work" is to transport the vehicle and payload a distance using the least amount of fuel. Is a big truck more efficient than a lightweight car? By a measure of ton-miles per gallon, it usually is. You can see that the simple concept of efficiency can easily become complex. Complexity leads to differences of opinion and arguments.

Effectiveness, on the other hand, describes the usefulness of the work output, which is not always its quantity. For example, let's say you drive forty miles to deliver four people to a location that is only twenty miles away, which takes you an hour and consumes one gallon of fuel. If you built the car, you would likely say the efficiency is quite good—forty miles per gallon, or even eighty passenger-miles per gallon. But if you were a passenger, the efficiency might not be so good—after all, you don't care about the others, and all you know is that it took one gallon to go twenty miles. How about effectiveness? In this case, the only thing you might care about is how long it took you to get there. How about the driver? Maybe he didn't want to go at all but had to deliver the passengers. He's not thrilled and his efficiency and effectiveness are both zero, unless they each paid for the transportation.

Let's evaluate another example. Say you are responsible for a city bus system and are told to increase the *efficiency*. What number will you use? Why not passenger miles per gallon? Makes sense, doesn't it? The easiest way to do that is to increase the number of people on each bus. Since the number of people who need to go to work every day is fixed (it is? are you sure?), let's just reduce the number of buses. As if by magic, the average number of people per bus will go up. Done—and you can take the rest of the day off! Not so fast—you misunderstood your boss. Your boss meant to say that you needed to increase the *effectiveness* of the bus system. Now how do you evaluate it? You first have to identify the objective of your customers, which is to travel between their homes and work in the least amount of time (and of course for the least cost). Then you have to measure the average time it takes for riders to go a given distance—not just on the bus, but from the time they leave their house until they walk into their office (average travel speed). And how are you going to increase the average speed? On the surface, this is easy: you can increase the number of buses so the wait at the bus stop is reduced, or you can also double the number of buses but have each bus stop at only half the number of stops.

So, in the example above, what is the logical result of the efficiency improvement compared to the effectiveness improvement? In the first case, the number of buses is reduced, and in the second, it is increased. Completely opposite! What is one to do? The lesson is that a superficial approach to *efficiency* improvement will quite often degrade the *effectiveness* of the system, leading to customer dissatisfaction and eventual failure.

The real objective is to balance efficiency and effectiveness in such a way that overall system performance is maximized. In the bus example, perhaps a detailed study of rider patterns will show that some buses could be express buses that skip stops, while smaller buses could be used for shorter distances. The idea is to reduce operating cost (increasing efficiency) while increasing passenger satisfaction (increasing effectiveness).

Consider this final example: Airline operating cost is dominated by fuel cost, so perhaps passenger miles per gallon is a good measure of efficiency. Effectiveness might be measured by the time it takes to go, let's say, from Detroit to Frankfurt. Airbus has proposed one approach to efficiency—the A-380, the largest passenger plane to date. By increasing the number of passengers, the passenger miles per gallon is increased by economies of scale (they didn't have to double the size of the aircraft to double the number of passengers). Boeing, on the other hand, is proposing the 787, the most efficient aircraft to date but perhaps no more efficient than the A-380 on a passenger-mile-per-gallon basis. The A-380 might be more efficient on an overall basis—fewer aircraft and crew are required to transport a given number of people. But a passenger may need to go to Frankfurt from Detroit, which the A-380 might not serve, partly because its sheer size doesn't allow it to fit at the gates and partly because there aren't enough passengers per day to justify a direct flight. As a result, passengers may need to fly first from Detroit to New York and then change planes to go on to Frankfurt. Or, would it be more desirable to spend slightly more to fly the smaller 787 nonstop from Detroit? Which do you think the average customer would prefer? I, for one, would take the nonstop every time, even if it costs a little more—it is simply a more *effective* way to get there.

Efficiency versus effectiveness: don't get caught in the efficiency trap; if the result is a reduction in effectiveness, you will likely lose in the long run.

Priorities: Importance versus Urgency

Someone once wisely said, "Important things are rarely urgent, and urgent things are rarely important." The idea is that if something is very important, it is unlikely that it just became known and therefore should have been addressed before it became urgent. And if something becomes urgent, it probably isn't important or it wouldn't have been put off until the last minute.

There are many people—and management seems to be especially prone to this attitude—who feel that only the "highest priority" items should be addressed at each moment, and these items are usually urgent things mistakenly identified as important. When one "highest priority" item is complete, the next "highest priority" item can then be addressed—but not until then. Of course, that almost guarantees nothing will be done, important or not, until it is urgent. Certainly not the way to run a company, although we seem to run a country that way.

So, how does one deal with priorities? It might be a good idea to separate important items from urgent ones. Does the task seem important only because it needs to be completed today? Or is the completion of the task truly important to the future of the company, your career, or you personally? Certainly if the fire alarm just sounded, getting out of the building is both important *and* urgent, but these occurrences are not the norm.

Another example: Let's say some design detail must be determined today or the tooling fabrication will continue with the design as it was. The design detail has to do with the logo configuration on the part. Is it urgent? Absolutely, since there is only the rest of the day to complete the task. Is it important? Probably not, since in this case the effect on the final product is very minor. At the same time, a quote is due next week that could double the dollar volume of the company. Important? Absolutely. Urgent? Not exactly—as long as there is time between now and the due date to accomplish the task. So what do you do today? Right now? Perhaps with five minutes of effort, you could finalize the logo and still have plenty of time to get the quote at least started. But the boss might disagree—the quote is the "highest priority," so drop everything else until it is done. That guarantees the logo on the part will never have the desired configuration. Or the boss might say, "What are you doing working on the quote—it's not due until next week. Get the logo configured since that has to be done today!" That might prevent the quote from getting out on time. Which to do? The following might offer some ideas on what to do about such a problem.

How do you come up with a list of priorities in the first place? The best way might be from the top down. What is the overall goal of the company (or you personally)? Is there a key project or product that fulfills that goal? What are the elements of that project? Break it down until the elements are a workable size; these can probably be considered high-priority items. And a completion date almost always accompanies each item. If you are responsible for the bracket design for a product, the high-priority item is not "bracket," or even "bracket design," but probably more like "the completion of a successful bracket design by January." Sounds like what some would call an objective or a milestone. It is defined, is measurable, and has a completion date attached.

Some would say this is easy—whatever has to be completed today is urgent. But is it? What about the task that has to be completed in a month? What if it depends on getting a quote from a supplier? You're going on vacation for a week, and if you get the quote request to the supplier, you will have a quote in hand when you return. Spending the thirty minutes to get the request for a quote out today will give you a week's advantage with no effort on your part (the supplier will be working while you are lying on the beach – life doesn't get better than that). So getting that RFQ out in the next thirty minutes could be both high priority *and* urgent, but at the least it is highly urgent. The reason is that it has a lot of "leverage"—thirty minutes of work today could speed up the project by a week.

One approach to determining priorities might be to list all the candidate activities—for the day, week, or year. Assign one number to the importance and another to the urgency for each task. I suggest making the numbers a geometric progression—each is double the previous. For example, "meeting Fred for lunch" (it's now 11:00 a.m.) might be urgent, but a simple phone call could cancel with no hard feelings, so it's very low on the priority scale. But maybe meeting Fred, who happens to be the CEO of a prospective employer, is crucial to your career. That places it very high on the priority scale.

Once you've assigned the importance and urgency numbers for each activity, multiply the two to get a third (priority) number. The activities with the highest five numbers get at least some of your time today. So what do you call these highest-rated activities?

Perhaps "productive" activities. Concentrate on the highest productivities. Here's a sample list:

	Priority	Urgency	Productivity
Meet Fred for lunch	1	8	8
Get PhD in 5 years	16	2	32
Finish report by tomorrow	4	4	16
Be home for kid's soccer game	8	4	32
Finish proposal for co. picnic by next month	2	2	4

So what should you spend time on today? Getting your PhD may rely on getting an application to a school. Or picking a school. Or figuring out what you want to do with the degree. It would be good to spend at least some time on that long-term priority. Next, arrange your day to be home in time for the soccer game—that game will never be played again, so to miss it will be to miss it forever. Then set aside enough time to make progress on tomorrow's report so it will be guaranteed to be done on time. Meet Fred for lunch? Maybe. You are probably going to take time out for lunch anyway, so it might not take any time at all. Go for it—but maybe call him and change the location to a place closer to your office. Think about the company picnic? Are you kidding? Let somebody else do it—don't waste even a second of your time on it.

So what might your boss say? "The report has to be finished by tomorrow—if I catch you spending a second working on anything else, you're fired!" Now what? Educate your boss on the benefits of rating priorities and urgencies to arrive at a list of productivities? Probably too late for that. That mentality is, I believe, one of the most destructive. Whether you can soldier on, getting most of what needs to be done finished on time without getting crossed up with management, is the real question. Over time you can only hope that management (everyone has "management," whether it's a first-line supervisor or the board of directors) will recognize your contribution. And perhaps even come around to your commonsense point of view!

Good luck—your career and your sanity may rely on your success in this endeavor.

The Natural Order of Things

I was once criticized for my slow solution to a problem. The boss said, "Why did you wait until last to try the solution that worked?" I replied, in a career-limiting way, "Good idea. In the future we will simply reverse the order we do things and do the last thing first."

Sometimes engineering is simple. But every myth and every seemingly irrational custom has some logical basis. So maybe we should think about doing the "last thing first." Or more rationally, we should put some effort into doing a series of things in the most logical order. As I said earlier, the easy decision is usually the wrong one. So let's take a minute and put the activities in a most logical order.

Answering the following questions might get you closer to an optimum order:

1. Is one activity quick and easy to accomplish? You might want to do that first if only just to get it out of the way.

2. Will one activity produce results that would be beneficial to another? Then it should be done first.

3. Based on your intuition, sometimes disguised as judgment, do you think a single activity could be "the

one" – the key to unraveling all the others? If so, you might want to do that one first—after all, you could get lucky.

4. Is there one activity that seems to be the "main man"? Resist starting on this one first, as it may take the time and resources that could be used trying several "lesser" tasks.

With the above information in hand, come up with a logical order. A good tool might be the Kepner-Tregoe analysis, described in the chapter "Quality Systems as a Business."

Here's an example: The yield of a process is too low—too many parts being scrapped. Upon investigation, it becomes obvious that the incoming parts have a problem, which is contributing to the reject rate. Also, a number of fixes could be applied to the process, which would help. But if it is assumed that the incoming parts will be improved (another group is working on that), should we concentrate solely on improvements that assume that incoming-part improvement? Or should fixes be made that will improve the situation immediately but be ineffective once the improved parts arrive? I suggest it might be logical to ignore the immediate fixes and concentrate on getting ready for the improved parts. Sure, it will at first look like no progress is being made, but when the improved parts arrive, maximum improvement will be instantly and seemingly magically realized.

So, take a quick look at the list of possible activities before you dive in—there will probably be a logical and commonsense order and it may not be the one predicted by rigorous analysis Who knows, you may indeed want to do the last thing first.

Epilogue and Final Thoughts

I hope the tools and discussions above have helped provide some understanding of the systems you will either use effectively or struggle with while attempting to accomplish your objectives. I drew the topics from those I have dealt with myself and where I have usually come to a conclusion different from the norm or what I was first told. The most important thing to be learned is to question existing norms if they run counter you your own common sense. I hope you have found some value in this book.

Some Random Thoughts

You can look at something but not see it

You can see something but not study it.

You can study something but not learn it.

You can learn something but not understand it.

You can understand something but not act on it.

People will remember you only for your actions

The easiest decision to make is usually the most difficult to implement.

Keep adding more and more simplicity—simplicity is the ultimate sophistication.

Important things are rarely urgent, and urgent things are rarely important.

Efficiency is often the enemy of effectiveness.

Don't ever start a question with, "Why don't you…?" You're likely to get an answer.

Every myth has a basis, however now irrational. Look for it to gain perspective.

Made in the USA
San Bernardino, CA
24 September 2016